The Abacus

The Abacus

Its history; its design; its possibilities in the modern world

PARRY MOON
Massachusetts Institute of Technology

GORDON AND BREACH SCIENCE PUBLISHERS
New York · London · Paris

Copyright © 1971 by
Gordon and Breach, Science Publishers, Inc.
440 Park Avenue South
New York, N.Y. 10016

Editorial office for the United Kingdom
Gordon and Breach, Science Publishers Ltd.
12 Bloomsbury Way
London W.C. 1

Editorial office for France
Gordon & Breach
7–9 rue Emile Dubois
Paris 14^e

Library of Congress catalog card number 72-143627 ISBN 0 677 01960 2 (cloth); 0 677 01965 3 (paper). All rights reserved. No part of this book may be reproduced or utilized in any form or by any means, electronic or mechanical, including photocopying, recording, or by any information storage and retrieval system, without permission in writing from the publishers. Printed in east Germany.

Contents

CHAPTER 1	Numbers	1
CHAPTER 2	The Counting Board and the Abacus	21
CHAPTER 3	Arithmetic	40
CHAPTER 4	Abacus Design	74
CHAPTER 5	The $(0, m)$ Abacus	101
CHAPTER 6	The (n, m) Abacus	126
CHAPTER 7	The (p, n, m) Abacus	141
APPENDIX A	Exercises in Abacus Arithmetic	150
APPENDIX B	Answers to Exercises of Appendix A	168
	Acknowledgments	175
	Index	177

"An educational toy in the West, a business tool in the Orient, the abacus holds an important place in man's technological history. In effect the primitive ancestor of the electronic computer of today, the abacus is a milestone in man's use of mathematics to master the physical world."

HARTLEY HOWE

"Abacus", in *Collier's Encyclopedia*, 1964

"The Abacus possesses a high respectability, arising from its great age, its widespread distribution, and its peculiar influence in the evolution of our modern system of arithmetic. In the Western lands of today it is used only in infant schools, and is intended to initiate the infant mind into the first mysteries of numbers... In India and over civilized Asia, however, the Abacus still holds its own; and in China and Japan the method of using it is peculiarly scientific."

C. G. KNOTT

"The Abacus", In *Napier Centenary Celebration Handbook* by E. M. Horsburgh, 1914

"There is (at least for me) a pleasure in rebelling against the enormous overcomplexity of modern life. 'Simplicity, simplicity, simplicity!' cried Henry David Thoreau... The abacus could hardly be simpler; yet it does exactly what it is supposed to do, with great efficiency, without electrical switches, vacuum tubes, transistors; without even a single lever or cogwheel... On the wall, near a giant computer, one sometimes sees an abacus behind glass, with the sign: IN CASE OF EMERGENCY, USE THIS. I for one take a perverse delight in operating a digital computer so simple that even the ancient Romans were able to manufacture it."

MARTIN GARDNER

In *The Japanese Abacus Explained* by Y. Yoshino (Dover Publications, New York, 1963)

CHAPTER 1

Numbers

In our modern age of giant computers, is it not ludicrous to consider anything as antiquated as the abacus? How silly can one get?

Actually, however, use of large electronic computers is justified only for the most complicated problems. The great bulk of routine calculations continues to be made, as in the past, by the pencil-and-paper arithmetic procedures that we learned in elementary school, supplemented in some cases by the slide rule, the logarithm table, and the desk calculator. Even if one belongs to the privileged class that has access to a multi-million dollar computer, he does not employ it to balance his check book or to calculate his income tax. To think that the electronic computer will eliminate the memorizing of the multiplication table is as nonsensical as that a TV set in every room will eliminate the need of learning to read. Arithmetic is a part of our cultural heritage—a part that tends to grow in importance as civilization becomes more complex.

Numerical computation is performed today in various ways:

1) Mental arithmetic.
2) The pencil-and-paper method.
3) Logarithm table, the slide rule.
4) The abacus.
5) Mechanical desk calculator.
6) Electronic desk calculator.
7) Electronic sequential computer.

Thanks to years of training in elementary schools, our daily lives include an astonishing amount of *mental arithmetic*, as when we estimate the total of our purchases in the chain store, the number of days of work before vacation, or the miles per gallon on an automobile trip. For slightly more complicated calculations of daily life, we resort to paper and pencil. The engineer also makes constant use of the *slide rule* (Fig. 1.1), which has the convenient property of giving him approximate results with just about the accuracy that he needs.

For higher accuracy, we have the ordinary desk calculator[1]*, consisting of a combination of mechanical gears, shafts, dials, etc., usually driven by an electric motor. In the present book, we mention these machines several times in comparison with the abacus. When the term "desk calculator" is used, we shall mean the familiar mechanical instrument rather than the electronic one. The electronic desk calculator (Figs. 1.2, 1.3) is a comparatively recent development. It is much faster than the mechanical model and can be made as versatile as desired.

The *electronic sequential computer*[2] (Fig. 1.4) is usually called a "digital computer". This name, however, is not distinctive since all the foregoing items except 3) are digital. The important aspect of such a machine is that it can be programmed to perform a whole sequence of operations automatically. This feature raises the sequential computer to a level vastly above that of the desk calculator, so that it can be programmed to play chess, compose music, or perform other feats usually considered to be human prerogatives.

Where does the abacus fit into this survey of computational methods? Apparently it could occupy a place between the ubiquitous mental arithmetic of the individual and his use of a desk calculator. There are frequent occasions where fairly extensive computations must be made which, however, hardly justify the purchase of a desk calculator. The abacus could be employed to good advantage for much of the work that is now done in 2) and much that is done in 5). Also, the abacus could be of distinct pedagogical value in the teaching of arithmetic in elementary schools. And it offers unique advantages in work with other number bases.

The abacus is a simple and inexpensive device. Its history is not well-documented, but the use of the abacus probably goes back at least 2000 years.[3] Some form of abacus was employed by the ancient Romans, and similar aids to calculation were common in Europe up to 1500 A.D. At about this date, a rivalry arose between those who used mechanical aids in calculation and those who did their mathematics with a pen. The advent of cheap paper (ca 1600) seems to have tipped the balance toward the latter.

In the orient, however, the abacus continued in popularity, so that today it is supreme in China, Japan, Russia, and many other countries. In modern

---

* References are given at the end of the chapter.

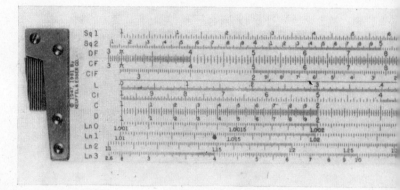

Figure 1.1 A

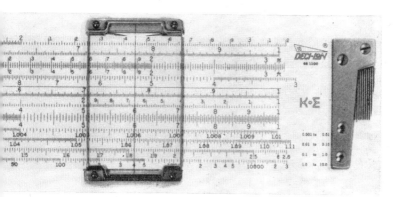

d Esser Co.)

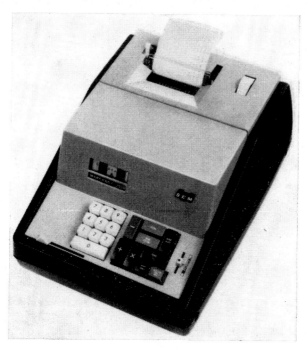

Figure 1.2 Electronic desk calculator. (Marchant)

Figure 1.3 Electronic desk calculator. (Friden)

Figure 1.4   Electronic sequential computer. (I.B.M.)

Japan, abacus instruction is an important part of the elementary-school curriculum. The abacus is employed regularly in stores, banks, and everywhere that a desk calculator is used in the United States. The Japan Chamber of Commerce and Industry provides examinations for licensing abacus operators.[4] Over a million such examinations are given each year, which indicates the importance of the abacus in Japanese life. The International Abacus Association is devoted to research on the abacus and to improvement in instructional methods. The Abacus Association of America has similar aims.

It is a surprising fact that the abacus in the hands of an expert is more rapid than a modern electrically driven desk calculator. A number of contests of this kind have been conducted. The first was held in Tokyo on November 12, 1946 under the auspices of the journal, *Stars and Stripes*. The abacus was definitely superior except in the multiplication of very large numbers. In fairness, however, it should be pointed out that to become an expert abacus operator, one must have a rigorous training that is on a quite different level from that required to become an expert desk-calculator operator.

In view of the undeniable importance of the abacus in the orient, it seems advisable for us in the west to reconsider the possibilities of the instrument. Beyond its possibilities for routine calculations, there appear to be two other considerations:

a) *Pedagogical.* Memorizing the multiplication table is a peculiarly abhorrent task for most children, and frequently it produces a permanent bias against mathematics. If the child is introduced to numbers via the abacus, however, he may glory in the new power that it gives him, just as he can be intrigued into learning the alphabet through that fascinating gadget, the typewriter. Later, he finds that the multiplication table is not too bad, the incentive being that it speeds up his abacus multiplication.

b) *Other number bases* There is a modern trend[5] toward the use of number bases other than the familiar Base-10. Such manipulations, however, introduce difficulties. Multiplication requires either learning a new multiplication table for each new base or the use of some kind of mechanical aid. The abacus appears to be an ideal aid, since it is easily modified for any number base.

Previous books[6] on the abacus have usually been practical instruction books dealing with a particular type of instrument. The present work aims at a broader treatment. We begin with a consideration of number systems (Chap. 1), followed by a survey of shortcuts in addition and multiplication (Chap. 3). These principles are then applied to the design of abaci for various number bases (Chap. 4). Finally, definite instructions are given for the operation of these abaci.

1.1 Counting

Counting, like talking, is an inherent characteristic of *Homo sapiens*. The most primitive tribes find a need to designate number of wives, children, or pigs, even though this number system may not exceed "one, two, many" or words to that effect. Counting on the fingers is almost universal. Thus, at an early stage in the development of any civilization, words and symbols are invented for what we call 1, 2, ... 9.

The symbol for *one* is usually a single stroke, either vertical or horizontal. Examples are shown in Fig. 1.5 (Egyptian), Fig. 1.6 (Attic), Fig. 1.7 (Roman), Fig. 1.8 (Chinese), and Fig. 1.9 (Hindu). The Sumerian and Babylonian numerals were made by pressing a stylus into a clay tablet.[7] The form of numerals is shown in Fig. 1.10. The Mayan numerals consist of dots and horizontal lines (Fig. 1.11).

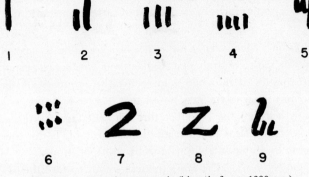

Figure 1.5   Egyptian numerals (hieratic form, 1000 B.C.)

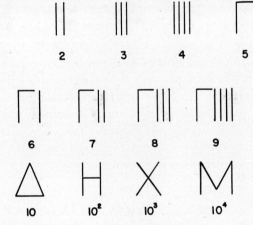

Figure 1.6   Attic numerals (1000 B.C.)

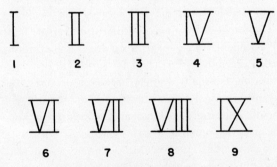

Figure 1.7   Roman numerals (1 A.D.)

NUMBERS

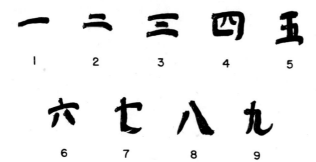

Figure 1.8 Chinese numerals (1 A.D.)

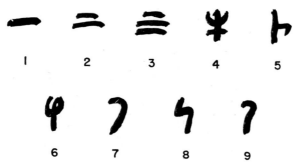

Figure 1.9 Hindu numerals (100 B.C.)

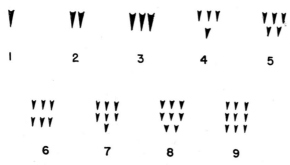

Figure 1.10 Babylonian numerals (2000 B.C.)

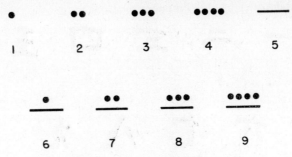

Figure 1.11   Mayan numerals (1 A.D.)

The basic notation for *two*, *three*, etc. consists in adding more strokes (Figs. 1.5 to 1.11). This cumbersome procedure was shortened in some cases by introducing a separate symbol for *five*. Thus a capital gamma Γ is used in Attic (Fig. 1.6), and a V is used in the Roman system (Fig. 1.7). A further refinement is introduced into the Chinese system (Fig. 1.8) and in the Hindu system (Fig. 1.9), where most integers from one to nine are represented by distinct symbols. The Hindu numerals were copied by the Arabs and were thus introduced into Spain, from whence they spread over Europe. An early set of European numerals is indicated in Fig 1.12.

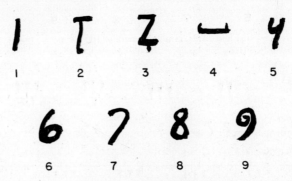

Figure 1.12   Gobar numerals.
Introduced into Europe about 1000 A.D.

What happens when primitive man needs to count above ten? One imagines, for instance, a shepherd who has more than ten sheep. He counts to ten on his fingers. He then moves a pebble or other counter, which he values at ten units, and again starts counting on the fingers. This process can be continued indefinitely. The final number of sheep is

$$N = 10a + c,$$

where a represents the number of pebbles and c is the number of fingers at the end of the counting process.

Note that the final result is still not in the concise form to which we are accustomed. Indeed, very few cultures ever succeeded in reaching the ultimate system which we call *positional*.[8] For instance, the Greek system (Fig. 1.13)

α	β	γ	δ	ϵ	ς	ζ	η	θ
1	2	3	4	5	6	7	8	9

ι	κ	λ	μ	ν	ξ	ο	π	ϙ
10	20	30	40	50	60	70	80	90

ρ	σ	τ	υ	φ	χ	ψ	ω	λ
100	200	300	400	500	600	700	800	900

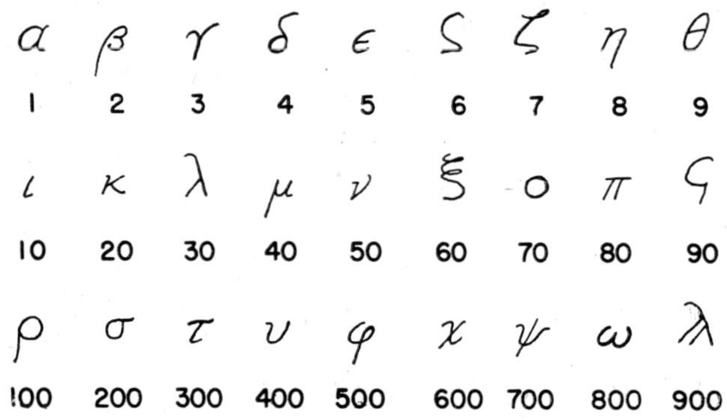

Figure 1.13 Greek numerals based on the Greek alphabet (500 B.C.)

represented numerals by letters of the Greek alphabet, plus three additional symbols (*digamma* for 6, *koppa* for 90, and *sampi* for 900). The ancient Greeks never realized that most of these symbols are quite superfluous. If we allow the value of a symbol to depend on its *position* with respect to the others, *we need only nine distinct symbols* in the decimal system, plus one more to indicate an empty space.

The Greek system (Fig. 1.13) is admirable in that no integer is represented by a complex, such as the nine marks representing 9 in the Babylonian system. The choice of the same symbols to represent numbers and letters of the alphabet, however, may lead to ambiguity. The Hebrew number system[5] has the same weaknesses as the Greek. In fact, these systems are in some respects inferior to the much older Greek system of Fig. 1.6.

Similar to the Attic system (Fig. 1.6) is the Roman system (Fig. 1.14), which again ignores the positional possibility. It may seem that we are overemphasizing the positional idea: after all, it is only a matter of notation;

Moon (0196)

isn't it? But here notation is everything. If we were confined to the Greek or the Roman system, not only would an ordinary multiplication or division be almost impossible but we would find a tremendous barrier to any advance in algebra or analysis.

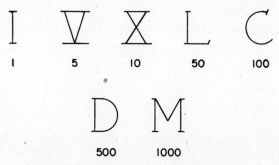

Figure 1.14 Roman symbols for large numbers.

When de Cordoba landed on the American continent in 1517, he found the ruins of the Mayan civilization. The Maya had long before developed a positional number system and had even invented a zero. The Babylonian system was also positional, though they seemed a little uncertain about the zero. Our positional number system appears to have been derived from that of the Hindus.[8]

Counting leads to the *positive integers* 1, 2, 3, ... The number concept, however, has been greatly extended during the past millenium. For integers, we may find it convenient to specify

$$\text{INTEGERS} \begin{cases} \text{Positive } 1, 2, 3, \ldots \\ \text{Non-negative } 0, 1, 2, 3, \ldots \\ \text{Negative } -1, -2, -3, \ldots \end{cases}$$

And of course there are many numbers that are not obtained by counting. *Real numbers* may be classified as follows:

$$\text{REAL NUMBERS} \begin{cases} \text{Rational} \begin{cases} \text{Integers} \\ \text{Fractions} \end{cases} \\ \text{Irrational} \begin{cases} \text{Algebraic} \\ \text{Transcendental} \end{cases} \end{cases}$$

Rational numbers include all the integers, and all the fractions obtained by dividing one integer by another; for instance, $\frac{1}{3}$ or $\frac{17}{32}$. The *irrational numbers* are all the other real numbers. They may be specified as *algebraic* if they are roots of polynomial equations with rational coefficients; otherwise they are *transcendental* (e.g., $\pi = 3.14159\cdots$)

1.2 Number Bases

We have seen how counting on the fingers developed quite independently in isolated tribes throughout the world and how this counting led inevitably to a certain periodicity in the number system. That is, at 11 the counting on the fingers started again, at 21 it began again, and so on. Thus the Base-10 number system arose independently in many parts of the world, even though most of these developments never reached the stage of a consistent notation, a symbol for zero, or a positional system.

In some cases, a primitive people counted on fingers and toes, leading to a Base-20 or vigesimal number system. Such a system was invented by the Maya in Central America and by the Basques in Europe. It still has echoes in the English language, where the word *score* occurs, and in the French word *quatre-vingt* (four twenties = 80). There was also a certain amount of counting on the fingers of one hand, which would lead to a Base-5 number system. There seems to be no historical example, however, of an independent Base-5 system. The familiar decimal system is, and always has been, the predominant one.

A noteworthy exception is the Base-60 number system of the Babylonians. It was developed in very early times. Sumerian records of this kind date back to 3000 B.C., and Babylonian mathematical tablets extend the use of Base-60 to about 300 B.C. Thus an important civilization utilized 60 as a number base for approximately three millenia.[9] The famous Greek astronomer Ptolemy (ca 200 A.D.) made all his calculations in this system[10], and we still have traces of it in our division of the circle and in our time divisions of the hour and minute.

Just a word about notation! In the Base-10 system, we use the word *digit* to refer to any of the number symbols from zero to nine. Since the word means "finger", it is not too good at best; and applied to other bases, it is completely inappropriate. To eliminate ambiguity, therefore, we can introduce a new word, *arithmos** (plural *arithmoi*) to designate the basic symbols for any base *b*.

---

* Greek ἀριθμος = number.

Similarly, the familiar term *decimal fraction* is inappropriate when applied to other than the decimal base. A general substitute for the term "decimal fraction", applicable to any base, is *meros* (from the Greek μέρος = part, fraction). These words will be used when necessary in the remainder of the book.

A number of thinkers have deplored the idea that humanity should be saddled forever with the Base-10 system just because some ignorant savages happened to count on their fingers. A scientific study should be made, they say, of the advantages and disadvantages of various number bases and a choice should be made of the most advantageous system. John Quincy Adams in a report to Congress, said[11] "Nature has no partiality for the number ten, and the attempt to shackle her freedom with it will forever prove abortive."

For, contrary to popular opinion, there is nothing magic about the Base-10 system. Analogous systems can be built for any base. Thus one might count in pairs, obtaining a Base-2 system; or one might use a base of 3, 4, etc. An interesting account of such arrangements is given in a book by John Leslie[12], published in 1820.

1.3 Theory

As noted previously, primitive man learned to count large numbers by repeatedly counting up to ten on his fingers and by using pebbles or other counters to represent tens, hundreds, etc. This process introduces a periodicity into the number system. Obviously, the period need not be ten. The traditional electronic digital computer, for instance, employs Base-2 because this base does not require the machine to discriminate among *ten* values of voltage or current but merely between an open circuit and a closed circuit or between positive and negative voltages.

For base b, a number x that is less than b and less than ten can be represented by the same symbol that we use in the Base-10 system. But for larger numbers, the representation is different for each base. Suppose, for example, that we have a group of objects, expressed by the number 73 in the Base-10 system and represented by the 73 unit counters of Fig.1.15. The 73 unit counters could be replaced by 36 counters of value 2, and one unit counter. Or they could be represented by 18 counters of value 4, plus a unit counter, or by 9 counters of value 8, plus a unit counter. Proceeding in this way, we finally arrive at one 64-counter plus one 8 counter plus one unit counter. Evidently the sum is 73 as it should be.

NUMBERS 13

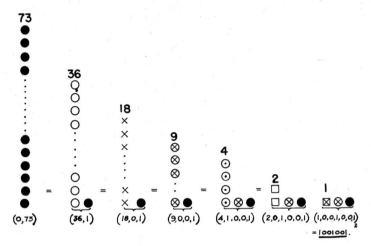

Figure 1.15 Representation of the number 73 in the binary system (Base-2).
● unit counter, 2^0; ○ 2-counter, 2^1; × 4-counter, 2^2; ⊗ 8-counter, 2^3;
⊙ 16-counter, 2^4; □ 32-counter, 2^5; ⊠ 64-counter, 2^6.

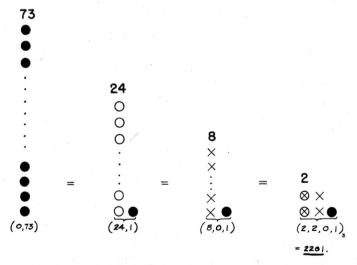

Figure 1.16 Representation of the number 73 in the ternary system (Base-3).
● unit counter, 3^0; ○ 3-counter, 3^1; × 9-counter, 3^2; ⊗ 27-counter, 3^3.

The counters in this example have values of 2^n, so the base is two. Thus the number, which is called 73 in the Base-10 system, may be written in the binary system as

$$(1,\ 0,\ 0,1,0,0,1).$$

Value   64 32 16 8 4 2 1

This may be shortened by omitting the commas and the parentheses, giving the number

1001001

in the binary system. Evidently, the binary representation is anything but compact.

Figure 1.16 shows the same process applied to the ternary number system (Base-3). The number 73 is here represented as *2201*. Obviously, the diagrams are quite unnecessary: to convert a Base-10 number to any base b, we merely divide successively by b. Thus for Base-2,

$$\begin{array}{r|l} 2 & 73 \\ \hline 2 & 36 + 1 \\ \hline 2 & 18 + 0 \\ \hline 2 & 9 + 0 \\ \hline 2 & 4 + 1 \\ \hline 2 & 2 + 0 \\ \hline & 1 + 0 \end{array}$$

The final 1 and the remainders, starting at the bottom and reading upward, give the number expressed in the binary system: *1 0 0 1 0 0 1*.

For Base-8, the same process expresses 73 as $(1,1,1)_8$ or *1 1 1*, where the first digit refers to 8^2, the second to 8^1, and the third to $8^0 = 1$:

$$\begin{array}{r|l} 8 & 73 \\ \hline 8 & 9 + 1 \\ \hline & 1 + 1 \end{array}$$

Similarly, for Base-12, 73 is expressed as $(6,1)_{12}$ or *61*; for Base-24, it is $(3,1)_{24}$ or *31*; and for Base-60, it is $(1,13)_{60}$. To prevent possible ambiguity, each number base should really have its own set of symbols. In the past, however, the familiar Arabic symbols have been employed wherever possible. With Base-12, for instance, one may write[13]

$$0, 1, 2, 3, 4, 5, 6, 7, 8, 9, \Delta, \varepsilon,$$

where the new symbols Δ and ε represent what we usually express as 10 and 11. The other digits are italicized to distinguish them from Base-10 symbols. The symbol *10* in the duodecimal system represents, of course, a dozen, while *100* represents a gross.

It is interesting to note how an increase in the base simplifies the writing of large numbers. This fact may have a practical bearing on the future writing of automobile registration numbers, telephone numbers, etc. As an example of simplification, some factorials are listed in Table 1.1. Note that 12! is expressed *exactly* by four digits in the duodecimal system but requires seven digits in the decimal system.

Table 1.1 Factorials

Number (Base-10)	Factorial (Base-10)	Factorial (Base-12)
1!	1	*1*
2!	2	*2*
3!	6	*6*
4!	24	*20*
5!	120	*Δ0*
6!	720	*500*
7!	5040	*2ε00*
8!	40,320	*1ε400*
9!	362,880	*156* × *10³*
10!	36,288 × 10²	*127* × *10⁴*
11!	399,168 × 10²	*1145* × *10⁴*
12!	4,790,016 × 10²	*1145* × *10⁵*

A whole number x may be written to any base b as follows:

$$x = a_n b^n + a_{n-1} b^{n-1} + \cdots + a_1 b + a_0, \qquad (1.1)$$

where $a_n, \ldots a_1, a_0$ are integers. This representation may be simplified by writing only the a's:

$$x = (a_n, a_{n-1}, \ldots, a_0)_b, \qquad (1.2)$$

or, if no ambiguity is thereby introduced, by merely writing

$$x = a_n a_{n-1} \cdots a_0. \tag{1.3}$$

For instance, if $b = 10$, $a_3 = 7$, $a_2 = 9$, $a_1 = 6$, $a_0 = 2$, we have

$$x = 7(10)^3 + 9(10)^2 + 6(10)^1 + 2(10)^0 = 7962.$$

A number x, given in the Base-10 system, is converted to base b by successive divisions by b, as shown in the previous examples. A number x, given to base b, is converted to Base-10 by use of Eq. (1.1). For example, the binary number $(1,1,0,1,0,1)_2 = $ *110101* is, according to Eq. (1.1),

$$x = 1(2)^5 + 1(2)^4 + 0(2)^3 + 1(2)^2 + 0(2)^1 + 1(2)^0$$

$$= 32 + 16 + 0 + 4 + 0 + 1 = 53$$

in the decimal system.

If x is not a whole number, the familiar idea of decimal fractions can be extended to any base, and

$$x = a_n b^n + a_{n-1} b^{n-1} + \cdots + a_1 b + a_0 + a_{-1} b^{-1} + \cdots + a_{-m} b^{-m} + \cdots \tag{1.4}$$

This representation is shortened by writing only the coefficients:

$$x = (a_n, \ldots a_1, a_0, a_{-1}, \ldots, a_{-m} \ldots)_b \tag{1.5}$$

The decimal point can also be introduced:

$$x = a_n \cdots a_0 \cdot a_{-1} \cdots a_{-m} \cdots \tag{1.6}$$

For $b = 8$, for instance, and $a_3 = 2$, $a_2 = 0$, $a_1 = 6$, $a_0 = 5$, $a_{-1} = 2$, $a_{-3} = 3$, Eq. (1.4) becomes

$$x = 2065.23$$

The choice of the best possible base b is not an easy one to decide. As b increases, the representation of large numbers becomes more compact but the multiplication table tends to become unwieldy. An important item to be considered is factorization of the base. A prime number for b is the worst possible choice because it has no factors. The choice of ten is almost as bad, since its only factors are 2 and 5 and the latter is not very helpful.

Experience of the human race shows that the *dozen* is very effective, since it is divisible into 2, 3, 4, and 6 parts. Thus in the Base-12 system, halves, thirds, quarters, and sixths are expressed very simply as meroi:

$$\text{One-half} = 0.6,$$
$$\text{One-third} = 0.4,$$
$$\text{One-fourth} = 0.3,$$
$$\text{One-sixth} = 0.2,$$
$$\text{One-twelfth} = 0.1.$$

This simplicity is a great help in division, and reacts on the entire system of Base-12 arithmetic.[14] Some other possible bases are listed in Table 1.2. Those having particularly helpful factors appear to be the bases 12, 24, 30, and 60.

Table 1.2 Factors of Numbers

Number	No. of factors	Factors
8	2	2, 4
10	2	2, 5
12	4	2, 3, 4, 6
16	3	2, 4, 8
18	4	2, 3, 6, 9
24	6	2, 3, 4, 6, 8, 12
30	6	2, 3, 5, 6, 10, 15
36	7	2, 3, 4, 6, 9, 12, 18
48	8	2, 3, 4, 6, 8, 12, 16, 24
60	10	2, 3, 4, 5, 6, 10, 12, 15, 20, 30
100	7	2, 4, 5, 10, 20, 25, 50

1.4 The Metric System

The metric system of weights and measures was developed at the time of the French revolution and was legally adopted in France in 1793. Most South American countries made it compulsory around the middle of the nineteenth century. The United States legalized it in 1866. Yet, after more than a century, it has not supplanted other units, even in France. W. R. Ingalls says[15]

Moon (0196)

The French Government was in 1906 complaining that it had been unable to suppress the use of old weights and measures; and I am under the impression that the same inability existed in 1939, as certainly it did in 1927 when Professor Kennelly examined the situation. Moreover ... for many practical purposes the metric system is so unsuitable that disregard for it is either sanctioned by law, or is condoned, or is dismissed as unworthy of attention. The idea that the metric system has come into sole use in any country, even under compulsion with penalty, is therefore a fallacy and a fiction.

Is opposition to the metric system merely an example of human inertia and stupidity or is there something inherently faulty in the metric proposal? Certainly the standardization of weights and measures is laudable. A basic trouble with the metric system, however, is its reification of Base-10. For scientific purposes, an immense range from the atomic to the astronomical is necessary; and the metric system with its unlimited powers of ten fulfills this requirement admirably. It is in its prosaic everyday aspects that the metric system is inadequate. Apparently one must be able to buy a *third* of a dozen eggs or a *quarter* pound of butter or measure a *sixteenth* of an inch; and none of these fits nicely into the Base-10 system, regardless of what fundamental units are adopted. The metric system emphasizes $\frac{1}{10}$ and $\frac{1}{100}$, neither of which is very helpful in everyday intercourse, while $\frac{1}{5}$ is almost as non-human as $\frac{1}{7}$ or $\frac{1}{11}$.

Herbert Spencer said[16]

Numeration by tens and multiples of ten has prevailed among civilized races from early times. What then has made them desert this mode of numeration in their tables of weights, measures, and values? They cannot have done this without a strong reason. The strong reason is conspicuous—the need for easy division into aliquot parts.

The metric system violates the natural human requirement of subdivision into 3, 4, 6, 8, ... parts. It is also inadequate for the subdivision of the circle, for the 24 hours in the day, the 12 months in the year, and the 32 points of the compass. All these requirements are met by changing the number base from 10 to 12, as has been pointed out repeatedly.[13] The advantages of the present metric system are retained and the disadvantages are overcome, merely by a change of base. Details are worked out in a book by Jean Essig.[17]

The metric system was introduced into this chapter merely as an example of the difficulties that the human race encounters when it slavishly adheres to the Base-10 system. Undoubtedly a better base could be found for *general use*. Of course, the practical obstacles in the way of changing the thinking of mankind from one base to another are formidable.

Actually, the obstacles may not be as hopelessly insurmountable as most people think. We have a vague feeling that our number system is immutable, has been with us since Creation, or at least since Sinai. The truth is that Arabic numerals were not generally accepted in Europe before 1500, and decimal fractions[18] were not invented until about 1600. Thus our present arithmetic is only 500 years old. It supplanted the Roman system which had been in use for 2000 years and the Babylonian system which lasted over 3000 years. Further changes are conceivable.

There is also the less-ambitious project of applying a *different number base* to each periodic phenomenon. Thus we might apply Base-7 to the 7 days in the week, Base-12 to the 12 semitones in the octave,[19] Base-60 to angle and time. This opens up a fascinating field for study.

As stated previously, each new number base requires a new addition table and a new multiplication table. Actually, the tables for the duodecimal system are more regular and more easily learned than for the decimal system. Thus if a world-wide change were made to Base-12, future children would have no added trouble on this score. For larger bases, however, the multiplication table becomes rather formidable.

A desk calculator, say for Base-60, would present an entirely new design and construction problem and would not be feasible unless there were a great demand for it. The abacus, on the other hand, is easily modified for any number base. Anyone can take a conventional soroban and change the frame so that it becomes an abacus for whatever base he wants. Thus the abacus seems to be the ideal tool for a person who feels an urge to study other number bases.

References

1. E. M. HORSBURGH, *Napier centenary celebration handbook*, Roy. Soc. of Edinburgh, (1914);
 A. GALLE, *Mathematische Instrumente*, B.G. Teubner, Leipzig, (1912);
 W. MEYER ZUR CAPELLEN, *Mathematische Instrumente*, Akad. Verlag, Leipzig, (1944);
 F. A. WILLERS, *Mathematische Maschinen und Instrumente*, Akad. Verlag, Berlin, (1951).
2. F. J. MURRAY, *Mathematical machines*, 2 vols., Columbia Univ. Press, (1961);
 W. C. IRWIN, *Digital computer principles*, D. Van Nostrand Co., Princeton, N.J., (1960);
 R. K. RICHARDS, *Arithmetic operations in digital computers*, D. Van Nostrand Co., Princeton, N.J., (1965).
3. D. E. SMITH, *History of mathematics*, vol. 2, chap. 3, Ginn and Co., Boston, (1925).

4. Japan Chamber of Commerce and Industry, *Soroban, the Japanese abacus*, Charles E. Tuttle Co., Rutland, Vt., (1967).
5. O. ORE, *Number theory and its history*, McGraw-Hill Book Co., New York, (1948).
6. T. KOJIMA, *The Japanese abacus, its use and theory*, Charles E. Tuttle Co., Rutland, Vt (1957);
    Advanced abacus, Tuttle, (1963);
    Y. YOSHINO, *The Japanese abacus explained*, Dover Publications, New York, (1963)
    Y. TANI, *The magic calculator*, Japan Publications Trading Co., Tokyo, (1964).
7. O. NEUGEBAUER, *Vorgriechische Mathematik*, Springer-Verlag, Berlin, (1934).
8. F. CAJORI, *A history of elementary mathematics*, Macmillan Co., New York, (1924)
    D. E. SMITH, *History of mathematics*, Ginn and Co., Boston, (1925);
    B. L. van der WAERDEN, *Science awakening*, P. Noordhoff, Groningen, (1954).
9. B. L. van der WAERDEN, op. cit., chap. 2.
10. PTOLEMY, *The Almagest*, Ency. Britannica, Chicago, (1952).
11. JOHN QUINCY ADAMS, "Report on weights and measures", Feb. 22, 1821.
12. J. LESLIE, *The philosophy of arithmetic*, Abernethy and Walker, Edinburgh, (1820).
13. F. E. ANDREWS, *New numbers*, Harcourt, Brace and World, New York, (1935);
    G. S. TERRY, *Duodecimal arithmetic*, Longmans, Green and Co., London, (1938).
14. T. LEECH, *Dozens vs. tens*, R. Hardwicke, London, (1866);
    G. W. COLLES, "The metric vs. the duodecimal system", p. 420, *A.S.M.E. Trans.*, 18 (1897);
    G. ELBROW, *The new English system of money, weights and measures, and of arithmetic* P. S. King and Son, London, (1913);
    K. MENNINGER, *Zahlwort und Ziffer*, F. Hirt, Breslau, (1934).
15. W. R. INGALLS, *Systems of weights and measures*, p. 14, Am. Inst. of Wts. and Meas New York, (1945).
16. HERBERT SPENCER, "Against the metric system", in *Works*, p. 142, D. Appleton and Co New York, (1897).
17. J. ESSIG, *Douze notre dix futur*, Dunod, Paris, (1955).
18. D. E. SMITH, *History of mathematics*, op. cit., vol. II, p. 240.
19. P. MOON, *A proposed musical notation*, p. 125, J. Franklin Inst., **253**, (1952);
    A scale for specifying frequency levels in octaves and semitones, p. 506, J. Acous. Soc of Am., **25**, (1953).

CHAPTER 2

The Counting Board and the Abacus

Mechanical aids to calculation are almost a necessity in business, and it is possible that some form of abacus was used by traders even in prehistoric times. The Phoenicians, the Egyptians, and the Greeks carried on a flourishing Mediterranean trade from very early times, and it is likely that they employed some kind of mechanical aid.

The early history of the abacus, however, is purely conjectural. Counting boards, if they existed, were probably crude affairs made of wood, which perished with the wood huts in which they were used. Classical painting and literature might be expected to give clues to the early history of the abacus.* Unfortunately, however, these aspects of life were the prerogatives of the chosen few. One could hardly expect the aristocratic Plato, for instance, to write a treatise on a device used by slaves and petty tradesmen!

2.1 The Salamis Counting Board[1]

Perhaps the oldest existing counting board was discovered on the island of Salamis, near Athens, about a century ago and is now in the National Museum at Athens. Its date is uncertain but has been estimated[2] as the fourth century B.C. It is a white marble slab 149 × 75 × 4.5 cm. On one face is a set of eleven parallel lines and some antique Greek numbers, Fig. 2.1. When the operator faced the long side of the slab, the lines of Fig. 2.1 were on his left. The center part of the board was blank, presumably for a pile of loose counters, and the right section contained additional lines used for fractions.

The counters, probably in the form of metal disks, were placed on the board *between lines*, the number of counters determining the particular digit.

---

* The only example of early art depicting a counting board seems to be a Greek vase Reference 2, p. 64) dating from the third century B.C. It shows Darius before his expedition against the Greeks (490 B.C.). The king's treasurer appears to be using a counting board.

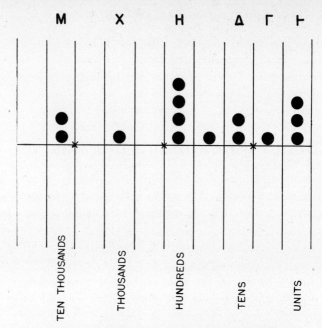

Figure 2.1 The Salamis counting board reading 21,478.

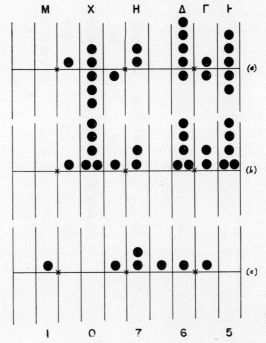

Figure 2.2 Addition 7248 + 3517 = 10,765 on the Salamis board.

THE COUNTING BOARD AND THE ABACUS 23

Besides the vertical spaces for units, tens, hundreds, etc., the Greeks also provided intermediate spaces for fives, fifties, etc. Thus counters were taken from the pile and placed in the proper spaces between lines to indicate the desired number. Figure 2.1 indicates 21,478.

Addition is easily effected on the Salamis board. Suppose, for example, that 7248 and 3517 are to be added, as in Fig. 2.2. The two numbers are set on the board, one above the horizontal division line and one below, Fig. 2.2a. The counters are then pushed together vertically, Fig. 2.2b, to give the sum.

This sum is then simplified, Fig. 2.2c. Starting at the right, we see 5 counters in the unit column. But these 5 counters are equal to one counter in the Γ-column; so we replace the 5 counters by 1 counter in the Γ-column, bringing the number in this column to 3. Since 2 of these 5-counters are equivalent to 1 counter in the 10-column, we replace 2 of the 5-counters by one counter in the Δ-column, bringing the total to 6. Five of these counters are then replaced by one in the 50-column. This procedure is continued, giving the sum 10,765 of Fig. 2.2c. Note that no mental work is required: the

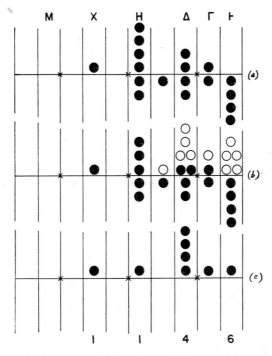

Figure 2.3   Subtraction 1425 − 279 = 1146 on the Salamis board.

process is very simple and entirely mechanical. After a little practice, the operator is able to add very quickly by this procedure, and various short-cuts expedite the work even more.

Subtraction on the Salamis board is carried out in a similar fashion (Fig. 2.3). The two numbers are represented on the board, just as in addition (Fig. 2.3a). The upper number is now readjusted so that it has at least as many counters in each column as the lower number. For example, the 4 unit counters in the lower number cannot be subtracted from the 0 unit counters in the upper number. So a counter in the 5-column is replaced by 5 counters in the unit column (Fig. 2.3b). In the drawing, these new counters are represented by white disks, though actually only one kind of counter is employed throughout. A similar procedure is used in the other columns (Fig. 2.3b)

The subtraction is then easily effected by taking the difference between the number of counters in upper and lower numbers. The result, 1146, is indicated in Fig. 2.3c. The counting board can also be used for multiplication, division, and the extraction of roots. Such operations, however, are relatively infrequent in commercial applications and will not be explained in this section.

2.2 Medieval Counting Boards

It is likely that counting boards were widely used in the ancient world and that the procedure of addition and subtraction was essentially that outlined in the preceding section. The device is sometimes called an *abacus* or a *line abacus*, from the Greek ἄβαξ, though the word *abacus* might preferably be reserved for the device with sliding counters permanently attached (Section 2.3).

The counting board continued in use with ever-increasing popularity through the middle ages until the 16th century, when it was gradually supplanted in Europe by our present paper-and-pencil calculations. The 16th century saw the publication of a great number of books on arithmetic which featured the counting board. Many pictures are still available showing the counting board in use, and the boards and counters are familiar museum pieces. Figure 2.4 gives a typical illustration from a book by Gregor Reisch published in 1504. It represents a European counting board (on the right) on which the numbers 1241 and 82 have been set up. The operator is supposed to be Pythagoras. On the left is Boethius, representing the "new arithmetic" done with a pen. In the background is Dame Arithmetic, who seems to be favoring the new method. This conclusion is emphasized by the worried expression on Pythagoras' face.

The literature of this period is filled with references to the counting board. A note dated 1556 says,[3]

Figure 2.4   Dame Arithmetic from Gregor Reisch, *Margarita Philosophica* (Strassbourg, 1504).

"Among Newe Yere's Guiftes gevon to the Quenis Maiestic (Mary Tudor) were by Mr. Surton a peire of tables, thre silver boxes for compters, and fourtie compters."

In France, the counters were called *jetons*,[4] and we have the popular verse[5]

> "Les courtisans sont des jetons,
> Leur valeur dépend de leur place;
> Dans la faveur, des millions,
> Et des zéros dans la disgrâce!"

THE ABACUS

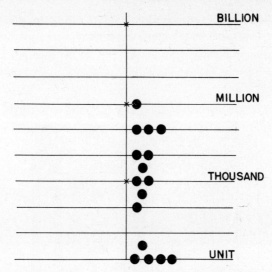

Figure 2.5 A medieval counting board with counters reading 1,327,609.

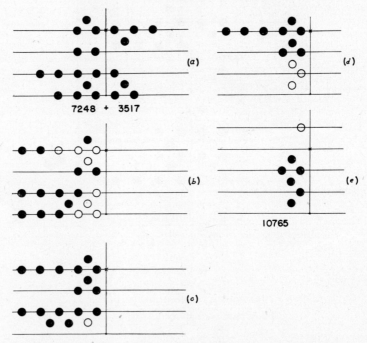

Figure 2.6 Addition on the counting board: 7248 + 3517 = 10,765.

THE COUNTING BOARD AND THE ABACUS 27

The standardized European counting board is surprisingly like the ancient Salamis board. There are two differences, however:

1) The diagram is turned through 90°; so in the European board, the numbers increase as we move away from the operator instead of increasing from right to left.

2) Counters are placed on the lines rather than between the lines. Only the 5, 50, ... are in the spaces. The arrangement is shown in Fig. 2.5.

Addition on the European board is indicated in Fig. 2.6. The two numbers are set on the board (a), one to the left, the other to the right of the vertical line. The counters are then moved together (b) to perform the addition. Finally, the result is simplified, as indicated in (c), (d), and (e). The sum is then read from the board in (e).

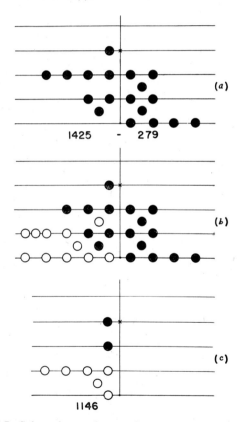

Figure 2.7 Subtraction on the counting board: 1425 − 279 = 1146.

Subtraction is carried out as with the Salamis board. The two numbers are set on the board, as shown in Fig. 2.7a The left one is then modified (b) so that it represents the same number but with at least as many counters on each line and space as occur on the right. Finally, the subtraction is made (c).

A popular method of *multiplication* in the middle ages was that of *duplation and mediation*. The theory is given in Chapter 3. Here we shall merely show how it was applied to the European counting board. The board shown in

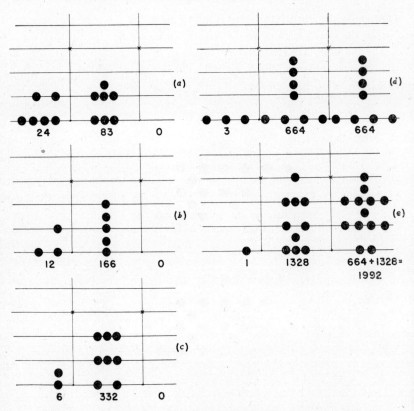

Figure 2.8 Multiplication on the counting board: $24 \times 83 = 1992$ by duplation and mediation.

Fig. 2.8 has three sections rather than two. The numbers to be multiplied are set on the first and second sections (a). The process then consists in successive *halving* of the first number and successive *doubling* of the second number. In halving, fractions are ignored. The procedure is continued until the left number is reduced to unity. The product is then obtained as the *sum of all the right-hand numbers associated with odd left-hand numbers*.

In the example 24 × 83, halving the left number and doubling the right gives 12 and 166, as shown in Fig. 2.8b. This process is continued to (e), where the numbers are 1 and 1328. There are only two cases where the left section contains odd numbers. The corresponding numbers (664 and 1328) in the center section are applied to the third section, giving the final result 1992.

This is a rather clumsy method of multiplication, though it does have the advantage of not requiring a knowledge of the multiplication table. The operator must be able merely to double and halve an arbitrary number. More conventional methods of multiplication are possible on the counting board. Division and extraction of roots are also feasible.

2.3 Other Abaci

Another form of abacus employs counters that are an integral part of the device and that slide in grooves or on rods. An early example is the Roman abacus[6] shown in Fig. 2.9. It consits of a metal plate with counters sliding in grooves. Ordinarily there are four counters in each of the long grooves and

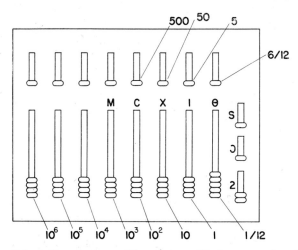

Figure 2.9 A Roman abacus. The counters slide in grooves in a bronze plate. The decimal system is used except for the fractions, which are in twelfths in accordance with Roman custom. Thus Θ are in *uncia* ($\frac{1}{12}$), S are in *semiuncia* ($\frac{1}{24}$), etc.

one counter in each of the short grooves. Since the counters cannot be removed to obtain a zero, a convention must be established regarding their zero position. Suppose that we take the position shown in Fig. 2.9 as zero. Then moving a counter away from the operator gives it a value. This value depends on the slot, in accordance with the decimal system. Thus each counter in the slot marked I has unit value, each counter in the X slot has value 10, etc. Each counter in the upper slots has a value five times that for the counter below. The 5 slots to the extreme right are for fractions and need not concern us here.

Comparison of the Roman abacus with the counting board shows that the two have many features in common, including the intermediate counters with values 5, 50, etc. The counting board is more flexible than the Roman abacus because the operator has an unlimited supply of counters to draw upon. On the other hand, the self-contained instrument is more compact, more portable, and is capable of more-rapid computation.

Some authorities believe that the Roman abacus was introduced into China early in the Christian era by traveling merchants.[7] The earliest mention of the abacus in Chinese literature, however, does not occur until the 12th century. The conventional Chinese instrument, or *suan-pan*, is still widely used. It is shown in Fig. 2.10. The arrangement is similar to that of the Roman abacus, but there are 5 unit beads on each rod instead of 4, and there are two 5-beads on each rod instead of one. This difference allows somewhat greater flexibility in the operation of the *suan-pan*, as compared with the Roman abacus.

The modern Japanese abacus or *soroban* is shown in Fig. 2.11. It was developed from the Chinese instrument. The sharp-edged beads are easy to manipulate, and the distance through which they move is small to allow high-speed operation. Until about 1930, an extra unit bead was included on each rod; but now the more compact instrument of Fig. 2.11 is employed throughout Japan.[8]

Finally, we have the Russian abacus (Fig. 2.12) with 10 beads per rod. Evidently this abacus differs from the counting boards and the previously described abaci by being a purely Base-10 instrument without any dependence on a sub-base 5. This necessitates the movement of a larger number of beads, but it is probably easier to learn than instruments having counters of value 5. While all the previous devices were operated horizontally, the Russian instrument can also be operated with the plane of the rods vertical. It is so used in the Soviet schools.[9]

THE COUNTING BOARD AND THE ABACUS

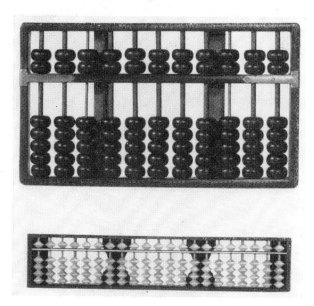

Figure 2.10 A Chinese abacus or *suan-pan* (above) and a modern Japanese abacus or *soroban* (below), giving a comparison in size.

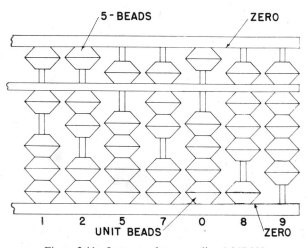

Figure 2.11 Japanese abacus reading 1,257,089.

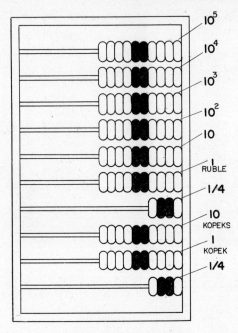

Figure 2.12 Russian abacus.

2.4 Abacus Operation[10]

Let us consider briefly the basic operation on an abacus with self-contained beads. Suppose that we illustrate the procedure by taking the Japanese *soroban*, Fig. 2.11. Other abaci will be considered in greater detail later. Here we merely want to present a rough idea of how an abacus is operated.

The abacus is first set at zero. It is customary to consider beads at zero if they are against the outside frame of the soroban. The instrument is tilted so that all beads slide toward the operator. The abacus is then set on a horizontal table or otherwise positioned so that the plane determined by the rods is horizontal. A movement of the forefinger along the 5-counters moves them upward (away from the operator) to their zero position.

We are now ready to set any number on the abacus by moving unit beads upward (away from the operator) and 5-beads downward (toward the operator). For instance, Fig. 2.13a shows the number 7248 set on the soroban.

Addition on the abacus is a little more difficult than on the counting board because of the strictly limited number of available beads. For instance, if we

THE COUNTING BOARD AND THE ABACUS

wanted to add 1 to the number set on the abacus in Fig. 2.13a, we would simply move upward the one available bead on the unit rod. But if we wanted to add 2, we would not have enough beads in the zero position. Thus we would have to think $2 = 10 - 8$, adding one bead on the 10-rod while sub-

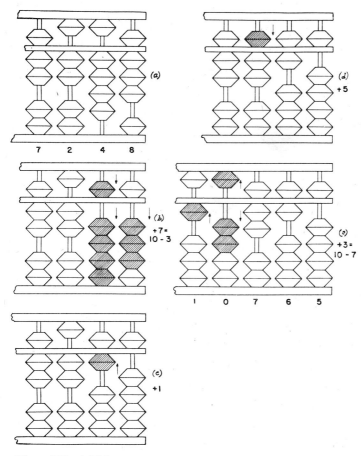

Figure 2.13 Addition $7248 + 3517 = 10{,}765$ on the Japanese abacus.

acting 8 (a 5-bead and 3 unit beads) on the unit rod. For each pair of digits, the soroban requires a unique movement or set of movements of the beads. Consider the example of addition that was used with the counting boards:

$$7248 + 3517.$$

The 7248 is first set on the abacus, Fig. 2.13a. On the unit rod, 7 must be added. But only one bead is available in the zero position. We therefore think 7 = 10 − 3, add one bead on the 10-rod and subtract 3 unit beads on the unit rod. In this particular arrangement, the operation is further complicated by the fact that no beads are available on the 10-rod. Thus the operator thinks 1 = 5 − 4 and moves a 5-counter downward (addition) and 4 unit counters downward (subtraction), as shown in (b). We next add the 1 of 3517 (Fig. 2.13c), the 5 (Fig. 2.13d), and finally the 3 (Fig. 2.13e). The sum 10,765 is then read from the abacus.

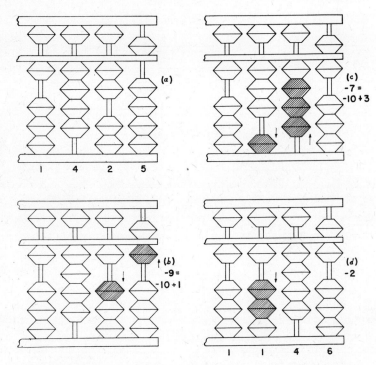

Figure 2.14   Subtraction 1425 − 279 = 1146 on the Japanese abacus.

We now take an example of subtraction:

$$1425 - 279.$$

The number 1425 is applied to the soroban, Fig. 2.14a. We next wish to subtract 9; but since this cannot be done directly because of the position of the

5-bead, we say $-9 = -10 + 1$. The operation is shown in (b). Similarly, the 7 of 279 cannot be subtracted directly but requires the relation $-7 = -10 + 3$ shown in (c). The final result (d) is 1146.

If the reader wishes to practice addition and subtraction on the soroban, he will find numerous problems in Appendix A. Answers are given in Appendix B. Or he may prefer to use the examples given in the Japanese books.[10]

2.5 Multiplication

The desk calculator is not a multiplying machine. It is not even an adding machine: it is a *counting machine* that utilizes the positional feature of our number system. The desk calculator does not know the multiplication table: it does not even know the addition table. For instance, in adding 2 and 6, it goes through all intermediate integers before it arrives at 8. And multiplication is performed by successive addition. Thus in multiplying 2718 by 7, the machine adds the multiplicand seven times, finally arriving at 19,026.

Like the desk calculator, the abacus is a *counting machine*. But with the abacus we have the advantage that the operator can know and use the multiplication table. Section 2.4 has shown that addition on the abacus is a purely mechanical process. Multiplication, on the contrary, is largely mental.

Multiplication by the conventional method learned in school consists of two parts:

a) Multiplication of all possible pairs of digits, using the memorized multiplication table,

b) Addition of the partial products.
In the product 73×47, for instance, one writes

$$\begin{array}{r} 73 \\ 47 \\ \hline 511 \\ 292 \\ \hline 3431. \end{array}$$

Here a) and b) are partly combined in the familiar method. In abacus multiplication, on the other hand, a) and b) are separated:

$$
\begin{array}{r}
7\dot{3} \\
\underline{47,} \\
3 \times 7 = 21 \\
7 \times 7 = 49 \\
3 \times 4 = 12 \\
7 \times 4 = \underline{28} \\
3431.
\end{array}
$$

The two cross-products (7×7 and 3×4) are written one place to the left of 21, while the 7×4 is written two places to the left. Part (b) then consists in pure addition of the four partial products.

Multiplication on the abacus is easy if one keeps in mind the elementary partial products and their positioning. Each partial product (a) is obtained mentally by the operator, who then applies this product to the abacus. Addition (b) is obtained automatically. In the above example, the operator

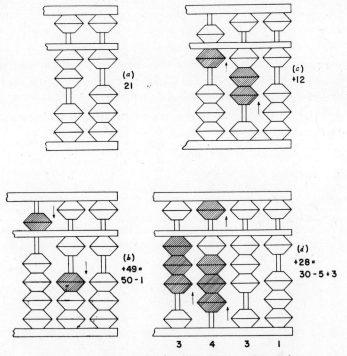

Figure 2.15 The product $73 \times 47 = 3431$ on the soroban.

knows that $3 \times 7 = 21$; and he represents this number on the abacus, Fig. 2.15a. The next product is $7 \times 7 = 49 = 50 - 1$, which is added to the previous number, as shown in Fig. 2.15b. We then introduce the remaining partial products (c) and (d) and read the final product 3431 from the abacus, Fig. 2.15d.

Figure 2.15 illustrates the condensed method of abacus multiplication, which is the exact analog of the paper-and-pencil method. The usual Japanese procedure[10] is somewhat more complicated, in that both multiplier and multiplicand are set on the instrument before starting work. Such a step may be advisable if the numbers are dictated orally. But for the usual case of written numbers, the additional step seems unnecessary and time-consuming.

The number of partial products increases rapidly as the number of digits increases. For example, a product 837×259 will require 9 partial products. We could write

$$
\begin{array}{rr}
 & 837 \\
 & 259 \\
\hline
7 \times 9 = & 63 \\
3 \times 9 = & 27 \\
7 \times 5 = & 35 \\
\hline
8 \times 9 = & 72 \\
7 \times 2 = & 14 \\
3 \times 5 = & 15 \\
\hline
8 \times 5 = & 40 \\
3 \times 2 = & 6 \\
8 \times 2 = & 16 \\
\hline
 & 216783 \\
\end{array}
$$

The corresponding manipulation on the abacus is left to the reader.

2.6 Division

Division has always been the most troublesome of the four arithmetic operations, and this continues to be true for abacus operations. The familiar process of *long division* is illustrated by dividing 3431 by 47:

$$
\begin{array}{r}
73 \\
47{\overline{\smash{\big)}\,3431}} \\
329 \\
\hline
141 \\
141 \\
\hline
\end{array}
$$

38   THE ABACUS

The 329 and 141 are obtained by a combination of products and may be dissected by writing

$$
\begin{array}{r}
73 \\
47 \overline{)\,3431} \\
\end{array}
$$

$$
\begin{array}{rl}
4\times 7 = & 28 \\
& \overline{63} \\
7\times 7 = & 49 \\
& \overline{141} \\
4\times 3 = & 12 \\
& \overline{21} \\
7\times 3 = & 21 \\
\end{array}
\Bigg\} 7 \quad \Bigg\} 3
$$

Here we take the products of integers, 4×7 and 7×7, separately and subtract them from 343. The next step consists in subtracting the products 4×3 and 7×3 from 141. This dissected procedure would not be advisable in the paper-and-pencil method, but it works nicely on the abacus.

Just as in multiplication, the abacus procedure in division follows directly from the paper-and-pencil analysis. The dividend is first placed on the instrument, preferably at the left end. A rod may be left vacant to the right of the dividend, and the quotient is placed to the right of the vacant rod.

The operator asks himself how many times 47 goes into 343 and decides on 7. The 7 is set on the seventh rod. The partial product $4 \times 7 = 28$ is next subtracted, and the product $7 \times 7 = 49$ is also subtracted. This leaves only 141 on the abacus. Evidently 141 is divisable by 47 three times. Thus a 3 is set in the quotient. We then proceed as before, subtracting the partial products $4 \times 3 = 12$ and $7 \times 3 = 21$. Finally the dividend disappears and only the quotient 73 remains.

As in multiplication, the standard Japanese procedure[10] is slightly more complicated, both dividend and divisor being set on the abacus. The detailed procedure is essentially the same in either case. The reader may use the Japanese method or the simplified method outlined above, whichever he prefers. Some examples for practice are given in Appendix A.

References

1. D. E. SMITH, *History of mathematics*, p. 162, vol. 2, Ginn and Co., Boston, (1925).
2. B. L. van der WAERDEN, *Science awakening*, p. 48, P. Noordhoff, Groningen, (1954).
3. K. MENNINGER, *Zahlwort und Ziffer*, p. 249, F. Hirt, Breslau, (1934).

4. F.P.BARNARD, *The casting counter and the counting-board*, Oxford Univ. Press, (1916);
   D.E.SMITH, *Computing jetons*, New York, (1921);
   F.YELDHAM, *The story of reckoning in the middle ages*, G.G.Herrap and Co., London, (1926);
   T.SNELLING, *A view of the origin, nature, and use of jettons or counters*, London, (1769);
   A.NAGL, *Die Rechenpfennige und die operative Arithmetik*, Vienna, (1888).
5. K.MENNINGER, op. cit., p. 278.
6. K.MENNINGER, p. 227.
7. D.E.SMITH and Y.MIKAMI, *A history of Japanese mathematics*, Open Court Pub. Co., Chicago, (1914).
8. C.G.KNOTT, "The abacus in its historic and scientific aspects", *Asiatic Soc. of Japan, Trans.*, **14**, (1886);
   The calculating machine of the East: the abacus, in E.M.Horsburgh, *Napier centenary celebration handbook*, p. 136, Roy. Soc. of Edinburgh, (1914).
9. Abacus, Collier's Encyclopedia, vol. 1, p. 4, Crowell-Collier Pub. Co., New York (1964).
10. T.KOJIMA, *The Japanese abacus, its use and theory*, Charles E.Tuttle Co., Rutland, Vt., (1957);
    Advanced abacus, Tuttle, (1963);
    Y.YOSHINO, *The Japanese abacus explained*, Dover Publications, New York, (1963);
    Y.TANI, *The magic calculator*, Japan Publications Trading Co., Tokyo, (1964);
    Japan Chamber of Commerce and Industry, *Soroban, the Japanese abacus*, Charles E.Tuttle Co., Rutland, Vt., (1967).

CHAPTER 3

Arithmetic

Chapter 2 has outlined the procedure in abacus addition, subtraction, multiplication, and division. In many cases, however, these basic procedures can be shortened and the operator can save time and effort by availing himself of various shortcuts. For instance, instead of multiplying by 9, he can multiply by 10 and subtract one times the multiplicand. This is a real timesaver if several 9's or 8's occur in the multiplier. With 999, for example, one simply adds three zeros to the multiplicand and subtracts the multiplicand.

A variety of such shortcuts is available.[1] The purpose of this chapter is to re-examine the subject of arithmetic, with particular emphasis on simplification. The question of memorizing multiplication tables also arises, since this may be particularly troublesome if one is using several number bases. Obviously, much of the material in this chapter is applicable to pencil-and-paper calculations and to the desk calculator as well as to the abacus.

3.1 Addition

Modern addition is merely a simplified and formalized *counting*. Instead of passing through all the intermediate digits, however, as in counting, we arrive immediately at the final digit by use of a memorized *addition table*. In adding 3 + 5, for instance, we know immediately that the result is 8, without counting 4, 5, 6, 7, 8.

For the Base-10 system, the addition table is

2	3	4	5	6	7	8	9	
4	5	6	7	8	9	10	11	2
	6	7	8	9	10	11	12	3
		8	9	10	11	12	13	4
			10	11	12	13	14	5
				12	13	14	15	6
					14	15	16	7
						16	17	8
							18	9

ARITHMETIC

We omit the row and column for unity as being trivial. Our addition table for Base-10 then contains only 36 distinct items. It is easily seen that for any base b, the number of distinct elements in the addition table is

$$n = \tfrac{1}{2}(b-2)(b-1). \tag{3.1}$$

That even such an obvious operation as addition cannot be taken for granted is shown by writing the addition table for a different base, say Base-8. *Here the numerals are italicized to emphasize that the base is not ten*, and or course $8 = 10, 9 = 11$, etc. The table is

2	3	4	5	6	7	
4	5	6	7	*10*	*11*	2
	6	7	*10*	*11*	*12*	3
		10	*11*	*12*	*13*	4
			12	*13*	*14*	5
				14	*15*	6
					16	7

The *abacus* is ideally suited for addition because it does not require the use of an addition table. An abacus for any base b is easily obtained by modification of a soroban. Once provided with the proper abacus, the operator can add and subtract very quickly, without memorizing an addition table and without mental effort.

The principal shortcut in addition consists in adding and subtracting an arbitrary number B. The addition of numbers x and y, with any number base, may be written

$$S = x + y = x + y + B - B = (B + x) - (B - y). \tag{3.2}$$

The scheme is helpful only if one of the numbers, say y, is not too different from a simple number like 10 or 100. Then B is taken as this simple number.

Suppose, for instance, that $x = 742$ and $y = 98$ with Base-10. We take $B = 100$, and think
$$742 + 98 = 842 - (100 - 98) = 840.$$
Similarly, taking $B = 1000$,
$$14983 + 989 = 15983 - 11 = 15972.$$
Or, with $B = 400$,
$$647 + 385 = 1047 - 15 = 1032.$$

The method is sometimes helpful in mental arithmetic. It is of limited applicability, but seems to be the only shortcut for addition. In multiplication, we shall find that the possibilities of shortcuts are much more promising.

3.2 Complements

A special case of the method of Section 3.1 employs *complements* with respect to base b. The result is, in effect, a reduction in the number of necessary digits, thus reducing the size of the addition table. With Base-10, such a step has little or no advantage, since everyone knows the complete table. But for other bases, the possibility of reducing the size of the addition table and the multiplication table should be considered.

Let $B = b$ in Eq. (3.2). Quantities x and y are digits from 1 to $(b - 1)$. The complements X and Y are defined as

$$X = b - x, \quad Y = b - y. \tag{3.3}$$

For the Base-10 system with x and y having values 1, 2, 3, 4, or 5, we use them directly; but if they are above 5, we use their complements. In this way, addition of digits above 5 is unnecessary and the addition table reduces to

	2	3	4	5	
4	5	6	7		2
	6	7	8		3
		8	9		4
			10		5

The corresponding table for Base-8 is

	2	3	4	
4	5	6		2
	6	7		3
		10		4

Evidently there are four possibilities in taking the sum of digits:

$$(1) \quad x \leqq 5 \text{ and } y \leqq 5. \tag{3.4}$$

Then $S = x + y$.

Example. $x = 2, y = 4,$
$S = 2 + 4 = 6.$

$(2) \quad x \leqq 5 \text{ but } y > 5.$

Then, from Eq. (3.2),
$$S = (b + x) - (b - y) = b + x - Y. \tag{3.5}$$

Example. $x = 3, y = 8$
$$S = 10 + 3 - 2 = 11.$$

(3) $x > 5$ but $y \leq 5$.

Then
$$S = b + y - X. \tag{3.6}$$

Example. $x = 7, \quad y = 4,$
$$S = 10 + 4 - 3 = 11.$$

(4) $x > 5$ and $y > 5$.

Then
$$S = 2b - (X + Y). \tag{3.7}$$

Example. $x = 7, \quad y = 8,$
$$S = 20 - (3 + 2) = 15.$$

Evidently the above method is usually more trouble than it is worth with Base-10. But for unfamiliar bases, used without an abacus, the reduction of the addition table to approximately one-quarter its normal size may be of distinct advantage. This advantage becomes even more pronounced with the multiplication table (Section 3.5).

3.3 Subtraction

As with addition, the abacus provides the ideal method for subtraction with any base. No table is required and the process is purely mechanical.

Subtraction introduces nothing really new. Corresponding to Eq. (3.2) we have
$$D = x - y = x - y + B - B = (x - B) + (B - y). \tag{3.8}$$
For example, if $x = 19723$ and $y = 976$, we take $B = 1000$, and
$$D = 19723 - 976 = 18723 + 24 = 18747.$$
Or, with $B = 100$ in a different example,
$$D = 1867 - 95 = 1767 + 5 = 1772.$$
Also, for $x = 19702$ and $y = 487$,
$$D = 19702 - 487 = 19202 + 13 = 19215$$
with $B = 500$.

3.4 The Multiplication Table

The familiar multiplication table for Base-10 is

2	3	4	5	6	7	8	9	
4	6	8	10	12	14	16	18	2
	9	12	15	18	21	24	27	3
		16	20	24	28	32	36	4
			25	30	35	40	45	5
				36	42	48	54	6
					49	56	63	7
						64	72	8
							81	9

Table 3.1 Number of Distinct Products in the Multiplication Table

Base b	Number of products n	n'
3	1	–
4	3	1
5	6	3
6	10	3
7	15	6
8	21	6
9	28	10
[10]	[36]	[10]
11	45	15
12	55	15
16	105	28
24	253	66
30	406	105
36	595	153
42	820	210
48	1081	276
60	1711	435
100	4851	1225

Today, everyone memorizes this table in elementary school and uses it throughout his life. In ancient and medieval times, such was not ordinarily the case. Authors of books on arithmetic, printed in considerable numbers in the 16th century, advised their readers to memorize. Metius (1635) says,

"Tabula Pitagorica, dieman wel vast in sijn memorie moet hebben." Ortega (1512) advocates, "... tabula bisogna sapere ad memoria como Ave Maria."

With each number base is associated a multiplication table. The number n of distinct products contained in this table is the same as for the addition table and is expressed by Eq. (3.1). As shown in Table 3.1, the number of products becomes very great for large b. It is doubtful if the Babylonians memorized their sexagesimal multiplication table with its 1711 entries!

Of course, the memorization of a multiplication table is not obligatory. For any given base b, we can write the table of products of integers and can refer to this table every time we take the product of digits. Such a procedure is time-consuming and requires that we always have our table with us. But in a way it is much easier than the use of a logarithm table because the multiplication table, referring as it does only to digits, is necessarily both *exact* and *compact*. Even for a large base b, it can be printed on one page, while a 10-place logarithm table may fill a thousand-page book.[2]

The usual multiplication table deals only with the product of *digits*, as $6 \times 7 = 42$. An interesting modification consists in making a table of products of *pairs of digits*. Such an arrangement is given by Leslie.[3] The table is too involved to be memorized but it gives much more information in a single entry than is obtained from the ordinary Base-10 table.

For example, consider the product

$$93319 \times 33674.$$

The ordinary method requires 25 products of digits and 30 additions. The Leslie table requires no multiplications and only 10 additions. Nine partial products are read from the table. The result is

```
            93319
            33674
           _____
             1406
             2442
              666
           _____
              684
             1188
              324
           _____
               57
               99
           27
           _____
         3142424006
```

The first step is to read the product 19 × 74 = 1406 from the table. The next is to read 33 × 74 = 2442, etc. The only peculiarity is that partial products are spaced two places apart instead of one. An extension to 3 digits is given by Crelle.[4] Advantages of the method are not sufficient to merit further consideration.

Another interesting proposal is the use of *quarter-squares*. We may write

$$(x + y)^2 = x^2 + 2xy + y^2,$$
$$(x - y)^2 = x^2 - 2xy + y^2.$$

Subtraction gives

$$(x + y)^2 - (x - y)^2 = 4xy.$$

Thus the product xy is

$$xy = \tfrac{1}{4}[(x + y)^2 - (x - y)^2]. \tag{3.9}$$

A table is prepared, therefore, of quarter-squares; and products xy are obtained by merely subtracting two values from the table. For example,

$$
\begin{array}{rl}
x = & 93319 \\
y = & 33674 \\ \hline
(x + y) = & 126993 \\
(x - y) = & 59645
\end{array}
\quad
\begin{array}{l}
\text{Values from Table} \\ \hline
4031805512 \\
889381506
\end{array}
$$

$$xy = 3142424006.$$

A table of this kind by A. Voisin[5] was printed in 1816, but the best-known table of quarter-squares is that of J. Blater.[6]

3.5 Condensed Tables

The next question is whether the condensed multiplication tables obtained by using complements are of any value. With base b, we wish to obtain the product xy, using the complements

$$X = b - x, \quad Y = b - y.$$

Evidently,

$$P = xy = xy + xb - xb = xb - x(b - y)$$

or

$$P = xy = xb - xY. \tag{3.10}$$

ARITHMETIC

For instance, for $b = 10$ and $Y = 1$,

$$1167 \times 9 = 1167(10) - 1167(1) = 11670 - 1167 = 10503.$$

The condensation of the multiplication table results from using only the smaller integers $1 \cdots 5$, taking complements for $6 \cdots 9$. Thus the range $2 \cdots 9$ of the Base-10 table is cut into two equal parts:

$$\overbrace{1|2345|6789|0}^{(b-2)}$$
$$\underbrace{}_{(b-2)/2}$$

Similarly, for Base-8,

$$\overbrace{1\ |\ 2\ \ 3\ \ 4\ |\ 5\ \ 6\ \ 7\ |\ 0}^{(b-2)}$$
$$\underbrace{}_{(b-2)/2}$$

and for Base-6,

$$\overbrace{1\ |\ 2\ \ \ \ 3\ |\ 4\ \ \ \ \ 5\ |\ 0}^{(b-2)}$$
$$\underbrace{}_{(b-2)/2}$$

For any base b that is an even number, a similar arrangement can be employed, and the multiplication table applies only to $(b-2)/2$ digits. The number of products in the table is then (for b an even number)

$$n' = \frac{b}{8}(b-2), \qquad (3.11)$$

which is approximately one-quarter the number n for the complete table. If b is an odd number, the corresponding equation is

$$n' = \tfrac{1}{8}(b-1)(b+1). \qquad (3.12)$$

Values are listed in Table 3.1.

In using the condensed multiplication table, we have four possible cases as in Section 3.2:

(1) $x \leqq b/2, \ y \leqq b/2$.

$$P = xy. \qquad (3.13)$$

Example. $x = 3$, $y = 4$, $b = 10$,

$$P = 3 \times 4 = 12.$$

(2) $\quad x \leq b/2, \quad y > b/2.$

$$P = xb - xY. \tag{3.10}$$

Example. $x = 5$, $y = 9$, $Y = 1$, $b = 10$,

$$P = 50 - 5 = 45.$$

(3) $\quad x > b/2, \quad y \leq b/2.$

$$P = yb - yX. \tag{3.14}$$

Example. $x = 8$, $y = 3$, $X = 2$, $b = 10$,

$$P = 30 - 6 = 24.$$

(4) $\quad x > b/2, \quad y > b/2.$

$$P = b^2 + XY - b(X + Y) \tag{3.15}$$

Example. $x = 8$, $y = 6$, $X = 2$, $Y = 4$, $b = 10$,

$$P = 100 + 8 - 10(6)$$

$$= 48.$$

The above limits are for b an even number. If b is odd, the limit $b/2$ is replaced by $(b + 1)/2$. The examples are, of course, trivial for $b = 10$. With an unfamiliar base, however, the condensed multiplication table may be an advantage.

3.6 Large Bases

Another method of reducing the size of the multiplication table employs a sub-base b' as well as the regular base b. This scheme is particularly applicable to large bases and was used by the ancient Babylonians with their Base-60 system.[7]

For Base-60 with $b' = 10$, the multiplication table is written for Base-10 in the usual way, supplemented by columns and rows for 10, 20, 30, 40, and 50. Any digit of the system is then either in the b'-part of the table or can be expressed as a sum. In this way, the 1711 items of the usual 60-table are reduced to 91 items.

ARITHMETIC

The method is applicable to any base b and any sub-base b'. As an illustration, consider $b = 12$ and $b' = 4$. The complete multiplication table is then

	2	3	4	8
2	4	6	8	14
3		9	10	20
4			14	28
8				54

which contains only 10 distinct items, as compared with 55 for the usual Base-12 table. The condensed table consists of three parts: the usual table for Base-4 (upper left), the products of 4 and 8 (lower right), and the cross products (upper right). For instance, the product

$$2 \times 3 = 6$$

is read directly. But the product 3×7 requires the addition of two values obtained from the table:

$$3 \times 7 = 3(4 + 3) = 3 \times 4 + 3 \times 3 = 10 + 9 = 19.$$

Similarly, the product $6 \times \varepsilon$ requires the addition of four readings:

$$6 \times \varepsilon = (4 + 2)(8 + 3) = 4 \times 8 + 4 \times 3 + 2 \times 8 + 2 \times 3$$
$$= 28 + 10 + 14 + 6 = 56.$$

Evidently, we have gained a simplified multiplication table, but at the expense of a more complicated procedure. For small bases, the disadvantages undoubtedly outweigh the advantages. But for bases of 30 or 60, this condensation scheme allows memorization of the multiplication table, which would be impracticable with the original.

In general, the condensed table is shown in Fig. 3.1, where

$$\xi = b/b'. \quad (3.16)$$

As in the previous example, there are three regions (I, II, III). According to Eq. (3.1), there are

$$\tfrac{1}{2}(b' - 2)(b' - 1)$$

4   Moon (0196)

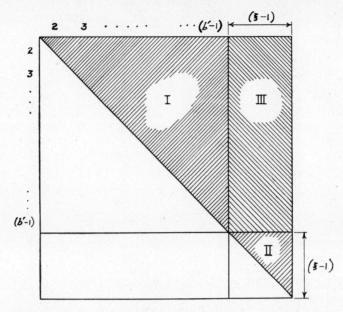

Figure 3.1 A schematic multiplication table for any bases b and b', with $\xi = b/b'$. The shaded areas contain numbers. By use of a sub-base b', multiplication tables for large b can be reduced in size.

distinct items in I,

$$\frac{\xi}{2}(\xi - 1)$$

items in II, and

$$(b' - 2)(\xi - 1)$$

items in III. Addition gives the total number of distinct products in the table:

$$N = \tfrac{1}{2}(b' + \xi - 3)(b' + \xi - 2), \qquad (3.17)$$

as compared with

$$N = \tfrac{1}{2}(b - 2)(b - 1) \qquad (3.1)$$

in the uncondensed table.

ARITHMETIC

For example, the usual multiplication table for Base-12 contains 55 products, but condensed tables have the following number of products, depending on the sub-base:

	$b' = 6$	4	3	2
ξ	2	3	4	6
$b' + \xi$	8	7	7	8
$b' + \xi - 3$	5	4	4	5
$b' + \xi - 2$	6	5	5	6
N	15	10	10	15

Similarly, for Base-30 with 406 distinct products,

	$b' = 15$	10	6	5	3
ξ	2	3	5	6	10
$b' + \xi$	17	13	11	11	13
$b' + \xi - 3$	14	10	8	8	10
$b' + \xi - 2$	15	11	9	9	11
N	105	55	36	36	55

For Base-60 with 1711 distinct products,

	$b' = 30$	20	15	12	10	6
ξ	2	3	4	5	6	10
$b' + \xi$	32	23	19	17	16	16
$b' + \xi - 3$	29	20	16	14	13	13
$b' + \xi - 2$	30	21	17	15	14	14
N	435	210	136	105	91	91

For arbitrary bases b and b', the given digits x and y may be written

$$\left.\begin{array}{l} x = A_1 b' + B_1, \\ y = A_2 b' + B_2, \end{array}\right\} \quad (3.18)$$

where A_i and B_i are integers or zero. The product is then

$$P = xy = (A_1 b' + B_1)(A_2 b' + B_2)$$
$$= A_1 A_2 (b')^2 + (A_1 B_2 + A_2 B_1) b' + B_1 B_2. \quad (3.19)$$

The product $B_1 B_2$ is read from part I of the table, $A_1 A_2$ is read from II, and $(A_1 B_2 + A_2 B_1)$ from III.

It should be realized that Eq. (3.19) gives only the product of *two digits*, just as with the usual multiplication table. A product of multi-digit numbers requires the usual addition of partial products.

3.7 Further Condensation

A further reduction in the number of products in the multiplication table is possible with the Base-10 system. In fact, the 36 distinct products of the customary table can be reduced to two: *doubling* and *halving*. Since both of these operations can be effected mentally, even with the largest multiplicands, the method has possibilities.

Multiplication of any multiplicand by 3 is obtained by adding the multiplicand and twice the multiplicand. Multiplication by 5 is obtained by adding a zero and dividing by two. The complete scheme is

$$3 = 2 + 1$$
$$4 = 5 - 1$$
$$6 = 5 + 1$$
$$7 = 5 + 2$$
$$8 = 10 - 2$$
$$9 = 10 - 1$$

In multiplying 3798×826 on the abacus, for instance, one would normally add 9 partial products:

$$\begin{array}{r} 3798 \\ \times\ 826 \\ \hline 48 \\ 54 \\ 42 \\ 18 \\ \hline 7596 \\ 64 \\ 72 \\ 56 \\ 24 \\ \hline 3137148. \end{array}$$

In obtaining the same product by the condensed method, one needs only 5 partial products. It is advisable as a first step to write down the double and half of the multiplicand:

$$3798 \times 2 = 7596,$$
$$3798 \times 5 = 18990.$$

Then, no matter how many digits are in the multiplier, all partial products consist of these two values. The abacus itself performs the additions and subtractions. In the above product, we first set 5 times the multiplicand on the abacus, then one times the multiplicand. This gives the 6 in the multiplier. Next add the 7596, which gives the 2. And finally represent the 8 by $+37980 - 7596$. The abacus then reads the final result, 3137148. The five partial products are

$$\begin{array}{r} 3798 \\ \times\ 826 \\ \hline 18990 \\ 3798 \\ \hline 7596 \\ \hline 37980 \\ -\ 7596 \\ \hline 3137148. \end{array}$$

Evidently, this method fits in nicely with the abacus and reduces the number of partial products, which tends to become excessive in the usual method of abacus multiplication (Section 2.5).

For $b < 10$, a similar procedure is possible. For the Base-8 system, we again double and halve, using factors of *2* and *4 = 10/2*:

$$3 = 2 + 1$$
$$5 = 4 + 1$$
$$6 = 4 + 2$$
$$7 = 10 - 1.$$

For Base-7, we need only the doubling:

$$3 = 2 + 1$$
$$4 = 2 + 2$$
$$5 = 10 - 2$$
$$6 = 10 - 1.$$

For Base-12, however, both doubling and halving are necessary, and the *9* even requires three partial products:

$$3 = 2 + 1$$
$$4 = 2 + 2$$
$$5 = 6 - 1$$
$$7 = 6 + 1$$
$$8 = 6 + 2$$
$$9 = 6 + 2 + 1$$
$$\Delta = 10 - 2$$
$$\varepsilon = 10 - 1.$$

The method is not helpful for large bases.

3.8 Duplation and Mediation

Closely associated with Section 3.7 is the ancient method of *duplation and mediation* mentioned in Section 2.2. This method was developed in ancient Egypt and was considered an important part of arithmetic in medieval Europe. The process consists in successive doubling of one of the numbers and successive halving of the other. For example, in multiplying 87 × 96, one writes

87	96
43	192
21	384
10	~~768~~
5	1536
2	~~3078~~
1	6144
	8352.

In halving, fractions are neglected. The final product is obtained by adding the numbers in the right-hand column, omitting all entries that correspond to even numbers in the left-hand column.[8]

This peculiar method of multiplication is really an application of the Base-2 system. The number 87 is transformed from the denary to the binary

ARITHMETIC 55

system, giving *1010111*. But

$$101011 = 1 + 1(2) + 1(2)^2 + 0(2)^3 + 1(2)^4 + 0(2)^5 + 1(2)^6.$$

Therefore,

$$87 \times 96 = [1 + 2 + (2)^2 + (2)^4 + (2)^6] 96$$
$$= 96 + 192 + 384 + 1536 + 6144 = 8352.$$

The numbers are recognized as the same as those obtained previously. Duplation and mediation is therefore merely a conventionalized form of this binary process. Practically, it has little to recommend it and is mentioned here only for its historical interest.

3.9 Multiplication

To have a table of products of all possible *integers* is, of course, entirely impracticable. What we do, therefore, is to have a multiplication table for *digits*. A product of multi-digit numbers is then obtained by finding the partial products of digits and adding them. Thus the multiplication process consists of two distinct operations:

a) Obtaining the partial products of digits,
b) Adding these partial products.

Multiplication on the abacus completely separates these two operations. Though b) is performed mechanically without the use of any addition table, a) must be done mentally, either by means of a memorized multiplication table or by reference to a printed table. The abacus by itself is completely incapable of performing operation a).

In multiplying an s-digit number by a t-digit number on the abacus, we have st partial products. For instance, the product 82×76 involves four partial products:

```
    82
  × 76
  ----
    12
    48
    14
    56
  ----
  6232
```

Of course, these partial products are not written: all four are introduced into the abacus, which then automatically reads 6232.

On the other hand, the ordinary method of multiplication, as taught in elementary schools, combines a) and b) to some extent so that the above example has only two partial products:

$$\begin{array}{r} 82 \\ \times\,76 \\ \hline 492 \\ 574 \\ \hline 6232 \end{array}$$

In general, the number of partial products is reduced by this procedure from st to the number of digits t in the multiplier.

There would be a great advantage if the abacus operator could train himself to produce mentally the product of any s-digit number by any digit. Perhaps the Trachtenberg method[9] offers some hope. There is also the possibility of using Napier rods for this purpose, as advocated by Domoryad.[10]

Ordinarily, then, the abacus product involves st partial products. But if the operator is using a condensed multiplication table, even more partial products are necessary. Thus condensed tables have distinct disadvantages. For example, if $b = 10$ and $b' = 5$, the condensed table is

	2	3	4	5
2	4	6	8	10
3		9	12	15
4			16	20
5				25

Corresponding to the product 82×76, we have

$$6 \times 2 = (5 + 1)\,2 = 10 + 2,$$
$$6 \times 8 = (5 + 1)(5 + 3) = 25 + 15 + 5 + 3,$$
$$7 \times 2 = (5 + 2)\,2 = 10 + 4,$$
$$7 \times 8 = (5 + 2)(5 + 3) = 25 + 15 + 10 + 6.$$

ARITHMETIC 57

The product can therefore be written

$$
\begin{array}{r}
82 \\
76 \\
\hline
10 \\
2 \\
\hline
25 \\
15 \\
5 \\
3 \\
\hline
10 \\
4 \\
\hline
25 \\
15 \\
10 \\
6 \\
\hline
6232
\end{array}
$$

Twelve partial products would be applied to the abacus, which would then read 6232. Of course, no one would think of using such a clumsy method for Base-10. For Base-60, however, a condensed multiplication table is not without merit.

3.10 The Gelosia Method

An interesting way of writing the partial products was called by the Italians *the gelosia method*,[11] from its fancied resemblance to the Venetian blind. An example is shown in Fig. 3.2. In each rectangular box is written the product

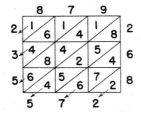

Figure 3.2 The gelosia method. In this example, 879 × 268 = 235572, which can be read directly from the diagram.

of two digits, with units below the diagonal and tens above the diagonal. The final product is written along the bottom, each digit being obtained by mentally adding all digits that appear between diagonal lines.[12]

Thus in the example, 879 × 268, shown in Fig. 3.2, we begin by writing 2, the final digit in the product, which appears in the lowest diagonal on the right. We next add $4 + 7 + 6 = 17$ for the next diagonal, writing 7 and carrying 1 for the next diagonal. The final result is 235572.

The original method was hopelessly slow, since it entailed the drawing of a rectangular network with diagonals and the insertion of a large number of digits. Once this preparation was made, however, the answer was immediately evident by inspection.

The gelosia method can be modernized by preparing a large number of rectangular plaques, each representing a product with diagonal separation of units and tens. These plaques are stored in a frame representing the multiplication table. To form any product, one merely takes the proper plaques from the frame and arranges them to form the desired product, as in Fig. 3.2. Evidently the plaques can be made equally well for any base. Experiments of this kind have been performed. Multiplication is delightfully free from mental strain, but unfortunately the process is still time-consuming. Perhaps some ingenious designer can develop a quick way of arranging the plaques.

3.11 Napier's Rods

The gelosia method is the basis for *Napier's rods* (Fig. 3.3). These rods were described by John Napier, the inventor of logarithms, in his book *Rabdologia*[13] published in 1617. The multiplication table is written on strips of cardboard or on the sides of square rods. The rods for a given multiplicand are assembled side by side (Fig. 3.4), and the partial products are read directly from the rods. Evidently this arrangement gives the multiplication table in a very convenient form.[14] Memorization of the table is unnecessary, and rods are easily constructed for any number base.

Napier's rods do not have the elegance of the original gelosia method. They do not give the final product by inspection: they give only the *partial products*, which must then be added to obtain the final result. The ideal way of adding these partial products is by abacus. Thus Napier's rods and an abacus make a delightful combination, particularly if one is working with an unfamiliar number base. No multiplication table is needed. Operation is quick and effortless.

ARITHMETIC

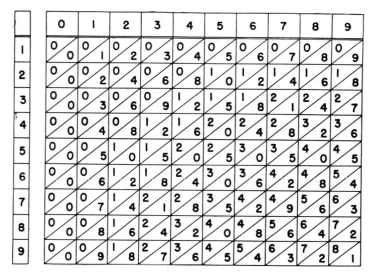

Figure 3.3 Napier's rods for Base-10. Each vertical strip represents a face of a rod. By placing the proper rods side by side, we can read any product.

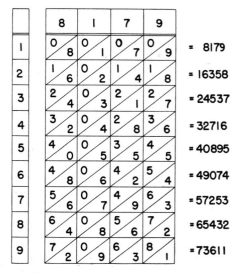

Figure 3.4 Napier's rods arranged for the multiplicand 8179.

In the example of Fig. 3.4, vertical strips labeled 8, 1, 7, 9 have been arranged for the multiplicand 8179. Digits of the multiplier are read from the separate strip on the extreme left. Each vertical strip contains the appropriate part of the multiplication table, units being printed below the diagonal line and tens above it. Products can be read directly from the rods, "carried" numbers being added mentally by summing diagonally. For instance, 2×8179 is $1/,/6 + 0/,/2 + 1/,/4 + 1/,/8$ or 16358. Similarly, 8×8179 is $6/,/4 + 0/ + 1$ from the next addition to the right, $/8 + 5/ + 1,/6 + 7/,2$ or 65432.

The arrangement of Fig. 3.4 gives the partial products for multiplying 8179 by any number. For example, in multiplying 8179 by 78, one reads the partial products 57253 and 65432 and adds:

$$\begin{array}{r} 57253 \\ \underline{65432} \\ 637962. \end{array}$$

Rods are easily made for any number base. An example for Base-12 is shown in Fig. 3.5.

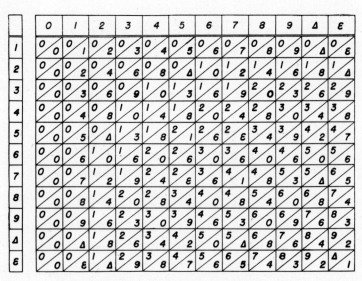

Figure 3.5 Napier's rods for Base-12.

3.12 Shortcuts

We shall now consider some obvious labor-saving schemes in multiplication. The idea of replacing 9 by (10 − 1) has already been mentioned. For instance, in multiplying a 10-digit number by 9, one makes 10 mental multiplications and a considerable number of mental additions:

$$\begin{array}{r} 55476 \quad 47763 \\ \times 9 \\ \hline 499288 \quad 29867. \end{array}$$

Employing the shortcut with the pencil-and-paper technique, one merely writes the number twice, shifted by a digit, and subtracts. Thus

$$\begin{array}{r} 554764 \quad 7763 \\ - \quad 55476 \quad 47763 \\ \hline 499288 \quad 29867. \end{array}$$

With a desk calculator, direct multiplication by 9 requires 9 revolutions of the gears, while subtraction requires only 1 revolution. With the abacus, the shortcut eliminates multiplication entirely and reduces the operation to a quick manipulation of the beads. In all cases, the shortcut is distinctly efficacious. In other words, *never multiply by 9*!

A similar procedure may be employed when multiplying by 8 or by 7. But here the advantage is not so marked because a mental multiplication by 2 or 3 is required.

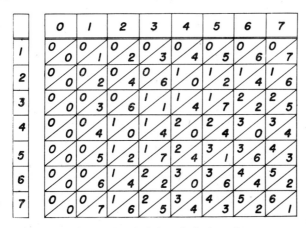

Figure 3.6 Napier's rods for Base-8.

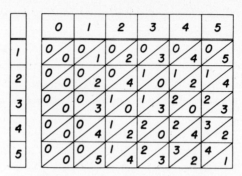

Figure 3.7   Napier's rods for Base-6.

The product with 11 is written directly by mentally adding successive integers. Trachtenberg[9] has developed a whole system of this kind, though the complexity of his rules militates against their use.

Another shortcut is to divide by 2 instead of multiplying by 5. The decimal point is moved, of course. The efficacy of this method rests on the fact that nearly everyone has developed considerable facility in doubling and halving. Thus the product of any number by 5 is written immediately by inspection. For example,

$$55476\ 47763 \times 5$$

is obviously obtained by adding a zero and dividing by 2, giving

$$277382\ 38815.$$

Comparative tests indicate that division by 2 can be accomplished in less than half the time for multiplication by 5. Our conclusion: *never multiply by* 5!

Many practical books are available on the subject of expediting multiplication.[1] These volumes consist largely of numerical examples, many of which are applicable only in very special cases. Analysis of the examples, however, shows that they are apparently classifiable under three heads depending on

1. Additive parts,
2. Factors,
3. Reciprocals.

1. *Additive parts*   This general scheme includes the familiar replacement of 9 by (10 − 1). It also includes the use of *complements*, where x is replaced by $(b - X)$ and y is replaced by $(b - Y)$.

The additive method 1. allows considerable flexibility. The multiplier is broken into any number of additive parts which are handled separately. Thus if $y = a + b + \cdots + k$, we can write

$$xy = xa + xb + \cdots + xk. \qquad (3.20)$$

If some of the pieces happen to be multiples, this fact can be used to advantage. As an example, take

$$79842 \times 3129.$$

We note that 12 and 9 are multiples of 3 and write

$$79842 \,[3000 + 4(3)\,10 + 3(3)]$$

or

$$\begin{aligned}
79842 \times 3000 &= 2395\ 26000 \\
239526 \times 40 &= 95\ 81040 \\
239526 \times 3 &= 7\ 18578 \\
\hline
&2498\ 25618.
\end{aligned}$$

A modification of this method consists in breaking both numbers into arbitrary parts. Let $x = a + \alpha$ and $y = c + \gamma$. Then

$$xy = (a + \alpha)(c + \gamma) = ac + a\gamma + c\alpha + \alpha\gamma. \qquad (3.21)$$

For example, take the product

$$71514 \times 81153$$

and arbitrarily select the first three digits for a and c:

$$a = 715 \times 10^2, \quad \alpha = 14;$$
$$c = 811 \times 10^2, \quad \gamma = 53.$$

The principal part of the product is

$$ac = 715(811)\,10^4,$$

to which are added the other terms, which are considerably smaller. Thus

$$\begin{aligned}
ac &= 715\,(811)\,10^4 = 57986\ 50000 \\
\alpha\gamma &= 715\ \,(53)\,10^2 = 37\ 89500 \\
c\alpha &= 811\ \,(14)\,10^2 = 11\ 35400 \\
\alpha\gamma &= 14 \times 53 = 742 \\
\hline
71514 &\times 81153 = 58035\ 75642.
\end{aligned}$$

The method is particularly helpful if, as is so often the case, we are interested in only a limited number of significant figures in the product. If the above numbers are accurate to only 5 figures, the product will be meaningless beyond 5 figures. Thus $\alpha\gamma$ is ignored and $a\gamma$ and $c\alpha$ are needed to only 2 figures. The product, therefore, reduces to

$$ac = 715\,(81.1)\,10^5 = 57987 \times 10^5$$
$$a\gamma = 7.15\,(5.3)\,10^5 = 38 \times 10^5$$
$$c\alpha = 8.11\,(1.4)\,10^5 = 11 \times 10^5$$
$$\overline{58036 \times 10^5.}$$

Therefore the product 71514×81153, correct to 5 figures, is 58036×10^5.

2. *Factors* If the multiplier can be factored into one-digit numbers, the product can be obtained without the addition of partial products. This principle sometimes allows a product to be obtained mentally for a problem that would normally required pencil and paper. For instance, 86×14 may be written

$$(86 \times 7)\,2 = (602)\,2 = 1204.$$

Or, even more simply, divide the multiplier by a factor and multiply the multiplicand by the same factor:

$$86 \times 14 = (86)\,2 \times 14/2 = 172 \times 7 = 1204.$$

This method is limited to multipliers that are factorable, and it does not seem to be any speedier than ordinary multiplication. In general, therefore, it is not recommended.

3. *Reciprocals* For Base-10, the most useful reciprocal was mentioned in Section 3.5, where instead of multiplying by 5 we divided by 2 with the proper shift of decimal point. Obviously, this procedure is valid also for multiples of 5.

This section has been confined to the Base-10 system. Most of the shortcuts can be extended, however, to other bases. As regards reciprocals, in fact other bases are usually better than Base-10 because of the notorious dearth of factors for 10. In Base-12, for instance, we have the factors *2, 3, 4, 6*. Thus instead of multiplying by *6*, we can divide by *2*; instead of multiplying by *4*, divide by *3*, etc. Larger bases may be even more advantageous in this respect.

3.13 Division

Division has always been the most troublesome operation in arithmetic. The Sumerians and Babylonians, as early as 2000 B.C., did their division by multiplying by the reciprocal. Extensive tables of reciprocals have been found, impressed on clay tablets.[15] The use of reciprocals was a powerful method because the base of the Babylonian number system was 60, which gives simple reciprocals for 2, 3, 4, 5, 6, 10, 12, 15, 20, and 30.

For the Base-10 system, on the other hand, the only factors are 2 and 5. Thus the principal aid to be expected from reciprocals is in division by 5, which can be replaced by multiplication by 2 (with shift of decimal point):

$$x/5 = 2x/10.$$

Related possibilities* are

$$x/25 = 4x/100$$

and

$$x/125 = 8x/1000.$$

These are examples of shortcut 3. of Section 3.12. For instance, with $a = 200$ and $c = 1$,

$$\frac{19879}{201} = \frac{198.79}{2} - \frac{198.79}{2}\left(\frac{1}{200}\right) + \frac{198.79}{2}\left(\frac{1}{200}\right)^2 - \cdots$$

$$= 99.395 - 0.497 \cdots + 0.002 \cdots = 98.900 \cdots$$

For 2., *factors*, simplification can sometimes be effected if the divisor is factorable. Thus if $y = ac$,

$$\frac{x}{y} = \frac{x}{ac} = \frac{(x/a)}{c}. \qquad (3.22)$$

For example,

$$\frac{1204}{14} = \frac{1204}{7 \times 2} = \frac{602}{7} = 86.$$

or

$$\frac{1314432}{168} = \frac{1314432}{8 \times 7 \times 3} = \frac{438144}{8 \times 7} = \frac{62592}{8} = 7824.$$

---

* See Section 3.14.

5 Moon (0196)

The method is particularly helpful if numerator and denominator contain common factors. For instance,

$$\frac{1204}{12} = \frac{43 \times 7 \times 4}{7 \times 2} = 86.$$

This is a kind of simplification that is often overlooked in practice.

For 3., *reciprocals*, the most useful is the replacement of division by 5 with multiplication by 2, as previously mentioned. Other number bases may give greater flexibility.

3.14 Reciprocals

The ancient Babylonians did their division by multiplying by a reciprocal; and extensive tables of reciprocals have been found, dating back to perhaps 1800 B.C. It is interesting to consider the possibility of this method for the modern world.

As is well known, the reciprocals of most integers are not expressible as decimal fractions with a finite number of digits. For instance, the reciprocal of 3 in the Base-10 system is 0.333 ⋯, the reciprocal of 7 is 0.14285714 ... Other bases may have advantages in this respect. Thus in Base-12, the reciprocal of *3* is exactly *0.4* and the reciprocal of *6* is *0.5*.

We now introduce a definition: A *regular number*[15] with respect to a given base b is defined as a positive integer such that its reciprocal is expressed exactly by a finite number of "digits". All other number are *irregular*.

What is the criterion by which one determines if a given number N is regular or irregular? Evidently the answer depends on the base. Here we need a fundamental theorem of arithmetic: *a positive integer can be written as a unique product of primes.*[16] Since b is an integer, it can therefore be expressed as a product of prime numbers. Suppose that

$$b = 2^k 3^l 5^m, \qquad (3.23)$$

where k, l, and m are positive integers. Usually bases do not contain primes above 5, so we shall limit our analysis to this case. The results, however, are easily extended to higher primes if such an extension is needed. If N is a positive integer, it is likewise expressible as a product of primes, or

$$N = 2^\alpha 3^\beta 5^\gamma 7^\delta \cdots \qquad (3.24)$$

where $\alpha, \beta, \gamma, \ldots$ are non-negative integers.

ARITHMETIC

We now prove the theorem: *The necessary and sufficient condition that a positive integer N be regular is that the same primes occur in Eqs. (3.23) and (3.24).*

1. Suppose that N is regular. Then its reciprocal $\bar{N}$ is a finite expansion:

$$\bar{N} = a_{-1}b^{-1} + a_{-2}b^{-2} + \cdots + a_{-p}b^{-p}$$

$$= \frac{1}{b^p}[a_{-1}b^{p-1} + a_{-2}b^{p-2} + \cdots + a_{-p}] = \frac{A}{b^p}. \quad (3.25)$$

Since A is a positive integer, it may be written as a product of primes. Thus

$$\bar{N} = \frac{2^t \cdot 3^u \cdot 5^v \cdot 7^w \cdots}{2^{kp}3^{lp}5^{mp}} = \frac{7^w \cdots}{2^{kp-t}3^{lp-u}5^{mp-v}}.$$

The reciprocal of $\bar{N}$ is therefore

$$N = 2^{kp-t}3^{lp-u}5^{mp-v}7^{-w}\cdots$$

But this can represent an *integer* only if all the exponents are non-negative. Thus the factors 7, 11, 13, ... cannot appear, and

$$N = 2^\alpha 3^\beta 5^\gamma, \quad (3.26)$$

and therefore, *a regular number is expressible in terms of the same primes as the base.*

2. Conversely, if Eq. (3.26) is true,

$$\bar{N} = \frac{1}{2^\alpha 3^\beta 5^\gamma}. \quad (3.27)$$

We wish to make the denominator a power of the base. Multiply numerator and denominator by $2^q 3^r 5^s$ to obtain

$$(2^\alpha 3^\beta 5^\gamma)(2^q 3^r 5^s) = b^p = 2^{kp}3^{lp}5^{mp}$$

or

$$2^{\alpha+q}3^{\beta+r}5^{\gamma+s} = 2^{kp}3^{lp}5^{mp}.$$

Therefore,

$$\begin{cases} \alpha + q = kp, \\ \beta + r = lp, \\ \gamma + s = mp. \end{cases}$$

or
$$\frac{\alpha + q}{k} = \frac{\beta + r}{l} = \frac{\gamma + s}{m} = p. \qquad (3.28)$$

Here we have three independent linear equations in the four unknowns p, q, r, s:

$$\begin{cases} q = kp - \alpha, \\ r = lp - \beta, \\ s = mp - \gamma. \end{cases}$$

Any one of these unknowns may be assigned arbitrarily, and the three equations then determine the other three unknowns. However, all four must be non-negative integers. Thus kp must equal or exceed the given α, with similar requirements for the second and third equations, or

$$\left. \begin{array}{l} p \geq \alpha/k, \\ p \geq \beta/l, \\ p \geq \gamma/m. \end{array} \right\} \qquad (3.29)$$

Equation (3.29) is satisfied by any integer above a certain minimum. Simplest manipulation is obtained by always taking this minimum value of p.

Returning now to Eq. (3.27), we multiply numerator and denominator by $2^q 3^r 5^s$, with the exponents satisfying Eq. (3.29):

$$\overline{N} = \frac{1}{2^\alpha 3^\beta 5^\gamma} \left(\frac{2^q 3^r 5^s}{2^q 3^r 5^s} \right) = \frac{2^q 3^r 5^s}{b^p}. \qquad (3.30)$$

The numerator is an integer, so it may be written

$$2^q 3^r 5^s = c_n b^n + c_{n-1} b^{n-1} + \cdots + c_1 b + c_0.$$

Substitution into Eq. (3.30) gives

$$\overline{N} = c_n b^{n-p} + c_{n-1} b^{n-p-1} + \cdots + c_1 b^{-(p-1)} + c_0 b^{-p}. \qquad (3.31)$$

Since $N > 1$, Eq. (3.31) represents a "decimal" fraction whose final digit is in the pth place after the decimal point. Thus we have proved that if N can be expressed as in Eq. (3.26), it is a regular number. We have also obtained the interesting result that the final digit in the reciprocal is in the pth "decimal" place, where p is obtained from Eq. (3.29).

ARITHMETIC   69

For the special case of $k = l = m = 1$, Eq. (3.29) reduces to

$$p \geq \alpha, \quad p \geq \beta, \quad p \geq \gamma,$$

resulting in the simple rule: *the final digit of $\bar{N}_r$ is in the α, β, or γ "decimal" place, whichever is the largest.* For example, if $k = l = m = 1$ and $\alpha = 3$, $\beta = 1, \gamma = 2$, then $p = \alpha = 3$. On the other hand, if the base exponents are not all unity, this rule must be modified by dividing by k, l, and m according to Eq. (3.29). Thus in Base-60, $k = 2$, $l = m = 1$. If $\alpha = 7$, $\beta = 2$, $\gamma = 3$, then

$$p \geq \tfrac{7}{2}, \quad p \geq \tfrac{2}{1}, \quad p \geq \tfrac{3}{1}$$

and the smallest integer that satisfies these relations is $p = 4$. The reciprocal will therefore have 4 "decimal" places.

Note that k, l, m are *positive* integers but α, β, γ are merely *non-negative* integers and may thus include zeros. For Base-60, for instance, $k = 2$, $l = m = 1 \neq 0$; but N may very well include zero exponents, as $\alpha = 8$, $\beta = \gamma = 0$. If the base does not need three primes, only the required ones are written. Thus for Base-10, we replace Eqs. (3.26) and (3.27) by

$$b = 2 \times 5, \quad N = 2^\alpha 5^\gamma.$$

Or, for Base-8,

$$b = 2^3, \quad N = 2^\alpha.$$

To find *all regular numbers* N_r, between given limits in the Base-10 system, one merely writes possible combinations of α and γ and determines the corresponding reciprocals.[15] For the range $1 \leq N_r < 10$, for instance, we have

α	γ	p	N_r	$\bar{N}_r$
0	0	0	1	1
0	1	1	5	0.2
1	0	1	2	0.5
2	0	2	4	0.25
3	0	3	8	0.125

This list is complete. The exponent p in Eq. (3.29) is the larger of the two integers α and γ and gives the number of decimal places in $\bar{N}_r$.

For a different range, $10^5 \leq N_r < 10^6$, there are 28 regular numbers, Table 3.2, as compared with the 5 for $1 \leq N_r < 10$. The spacing is non-uniform and varies from approximately 2 percent to 20 percent of N.

The fairly close spacing between regular numbers suggests the possibility of calculating reciprocals of *irregular* numbers by use of Table 3.2. Let any number N be expressed as

$$N = N_r + \Delta, \tag{3.32}$$

Table 3.2 Reciprocals of Regular Numbers for Base-10 and $10^5 \leq N_r < 10^6$

N_r	$\bar{N}_r$	N_r	$\bar{N}_r$
100000	1.0×10^{-5}	320000	0.3125×10^{-5}
102400	0.9765625	327680	0.30517578125
125000	0.8	390625	0.256
128000	0.78125	400000	0.25
131072	0.76293945312 5	409600	0.244140625
156250	0.64	500000	0.2
160000	0.625	512000	0.1953125
163840	0.610351562 5	524288	0.19073486328125
200000	0.5	625000	0.16
204800	0.48828125	640000	0.15625
250000	0.4	655360	0.15258789062 5
256000	0.390625	781250	0.128
262144	0.3814697265625	800000	0.125
312500	0.32	819200	0.1220703125

where N_r is a tabulated value. Then the reciprocal is

$$\bar{N} = (N_r + \Delta)^{-1} \doteq \bar{N}_r(1 + \Delta\bar{N}_r)^{-1}$$

$$= \bar{N}_r[1 - (\Delta N_r) + (\Delta N_r)^2 - (\Delta N_r)^3 + \cdots] \tag{3.33}$$

Since $(\Delta N_r) \ll 1$, high accuracy in its determination is unnecessary and the higher terms in Eq. (3.33) can often be obtained expeditiously on a slide rule.

For example, what is the reciprocal of 5.05137, correct to 5 significant figures? Take $N_r = 5$ and $\Delta = 0.05137$.
Then $\bar{N}r = 0.2$ and

$$\Delta \bar{N}_r = 0.01027,$$

$$(\Delta \bar{N}_r)^2 = 0.00010.$$

According to Eq. (3.33),

$$\bar{N} = 0.2\,[1 - 0.0103 + 0.0001] = 0.19796\ldots$$

ARITHMETIC 71

The possibilities of this method are greatly enhanced if one is working in other number bases. With Base-12, for instance, the regular numbers are more closely spaced than with Base-10, resulting in easier use of Eq. (3.33).

3.15 Factors

We have seen that breaking a number into its prime factors may be helpful in both multiplication and division. Such an operation, however, is not advisable in most cases because it may be more time-consuming than the multiplication or division that it is supposed to facilitate.

Fortunately, the multiplication table exhibits regularities that are an aid in factoring. For Base-10, we have regularities in the last digit of the product N, as shown in Table 3.3. For instance, if the final digit is an even number,

Table 3.3 Factors for Base-10

Final digit	Always divisible by	Sometimes divisible by
1	–	1, 3, 7, 9
2	2	3, 4, 6, 7, 8, 9
3	3	7, 9
4	2	3, 4, 6, 7, 8, 9
5	5	3, 7, 9
6	2	3, 4, 6, 7, 8, 9
7	–	3, 9
8	2	3, 4, 6, 7, 8, 9
9	–	3, 7
0	2, 10	4, 6, 8

N is always divisible by 2; if the final digit is 5, N is always divisible by 5. There are also other possibilities. If N ends in 7, Table 3.3 shows that there is no unique factor. But examination of the multiplication table yields the products $1 \times 7 = 7$ and $3 \times 9 = 27$, so 3, 7, 9 are possible factors. Similarly, if N ends in 9, there is no necessary factor; but $1 \times 9 = 9$, $3 \times 3 = 9$, and $7 \times 7 = 49$ show that 3, 7, and 9 are again possible factors.

As an example, consider
$$653037/2541.$$

Long division can be used in the familiar way. But if there are common factors in numerator and denominator, their use may expedite matters. Reference to Table 3.3

shows no necessary factors, though the third column suggests the possibility of 3 as a common factor. We immediately see that 3 is indeed a common factor. The table then suggests other possibilities and we write

$$\frac{653037}{2541} = \frac{217679}{847} = \frac{31097}{121} = \frac{2827}{11} = 257.$$

Similar tables can be prepared for other bases. With Base-12 or Base-30, for instance, the multiplication table exhibits much more regularity[17] than for Base-10. Thus factoring in these bases is generally easier than in the decimal system.

3.16 Summary

The chapter makes a cursory survery of arithmetic, particularly with the hope of finding time-saving devices in multiplication. Some of the topics turn out to be of little practical value, though their consideration may be of general interest. Let us select the methods that seem most promising.

1) For *addition*, the abacus has great advantages, especially with non-decimal bases for which desk calculators are not available. Addition on the abacus is purely mechanical and requires no memorization of addition tables.

2) For *multiplication*, the abacus operator must mentally supply all products of integers, the abacus being used merely to add the partial products. This means that the operator must have thoroughly memorized the multiplication table or must constantly refer to a table. For an unfamiliar base, the best solution of the problem appears to be the use of Napier's rods (Section 3.11) to give the partial products, with an abacus to add them.

3) For *division*, the combination of Napier's rods and abacus is again desirable. The use of *reciprocals* (Section 3.14) also seems promising.

In all cases, the operator should be on the lookout for shortcuts, such as use of complements, factors, double-and-halve, etc.

References

1. L. MEYERS, *High-speed mathematics*, D. Van Nostrand Co., Princeton, N.J., (1947);
   D. E. SMITH, *History of mathematics*, vol. 2, chap. 2, Ginn and Co., Boston, (1925);
   U. CASSINA, *Calcolo numerico*, N. Zanichelli, Bologna, (1928);
   J. F. BROWN, *Numbers and how to use them*, J. F. Brown, Fitchburg, Mass., (1892);
   M. C. VOLPEL, *Concepts and methods of arithmetic*, p. 119, Dover Publications, New York, (1964).

ARITHMETIC 73

2. J. PETERS, *Zehnstellige Logarithmentafel*, F. Ungar Pub. Co., New York, (1957).
3. J. LESLIE, *The philosophy of arithmetic*, p. 245, Abenethy and Walker, Edinburgh, (1820);
   H. ZIMMERMAN, *Calculating tables*, W. Ernst u. Sohn, Berlin, (1904).
4. A. L. CRELLE, *Calculating tables*, Products to 1000 × 1000, B. Westermann and Co., New York, (1898).
5. A. VOISIN, *Tables des multiplications, ou logarithmes des nombres entiers*, Paris, (1816); see also Leslie, op. cit., p. 249.
6. J. BLATER, *Table des quarts de carrés*, 1 to 200,000. Gauthier-Villars, Paris, (1888).
7. O. NEUGEBAUER, "Sexagesimalsystem und babylonische Bruchrechnung", *Quellen und Studien*, Springer-Verlag, Berlin, (1931), BI, pp. 183, 452, 458; p. 199, BII (1933); "Mathematische Keilschrifttexte", *Quellen und Studien*, A3, (1935);
   F. THUREAU-DANGIN, *Textes mathematiques babyloniens*, Leiden, (1938).
8. O. ORE, *Number theory and its history*, p. 38, McGraw-Hill Book Co., New York, (1948);
   D. E. SMITH, op. cit., p. 106;
   K. MENNINGER, *Zahlwort und Ziffer*, p. 270, F. Hirt, Breslau, (1934).
9. A. CUTLER and R. MCSHANE, *The Trachtenberg speed system of basic mathematics*, Doubleday and Co., Garden City, N. Y., (1960).
10. A. P. DOMORYAD, *Mathematical games and pastimes*, p. 47, Macmillan Co., New York, (1964).
11. French *jalousie*. Pacioli, in his famous arithmetic, *Suma* (1494), says, "Gelosia intendiamo quelle graticelle ch si costumono mettere ale finestre de la case dove habitano done acio no si possino facilme e vedere o altri religiosi. Diche molto abonda la excelsa cita de uinegia [Venice]."
12. D. E. SMITH, op. cit., vol. 2, p. 114;
    K. MENNINGER, *Zahlwort und Ziffer*, p. 349, F. Hirt, Breslau, (1934).
13. J. NAPIER, *Rabdologia*, Edinburgh, (1617); Leyden, (1626);
    W. LEYBOURN, *The art of numbering by speaking rods: vulgarly termed Nepeir's bones*, London, (1667).
14. E. M. HORSBURGH, *Napier centenary celebration handbook*, Roy. Soc. of Edinburgh, (1914);
    D. E. SMITH, op. cit., p. 202;
    M. F. WILLERDING, *Mathematical concepts, a historical approach*, p. 46, Prindle, Weber, and Schmidt, Boston, (1967);
    J. L. COOLIDGE, *The mathematics of great amateurs*, p. 82, Oxford Univ. Press, (1949);
    D. E. SMITH, *A source book in mathematics*, p. 182, McGraw-Hill Book Co., New York, (1929);
    K. MENNINGER, *Number words and number symbols*, p. 444, M.I.T. Press, Cambridge, Mass., (1969).
15. O. NEUGEBAUER, *Vorgriechische Mathematik*, Springer-Verlag, Berlin, (1934).
16. J. E. SHOCKLEY, *Introduction to number theory*, p. 17, Holt, Rinehart, and Winston, New York, (1967);
    O. ORE, op. cit., p. 4.
17. F. E. ANDREWS, *New numbers*, Harcourt, Brace and World, New York, (1935);
    G. S. TERRY, *Duodecimal arithmetic*, Longmans, Green and Co., London, (1938).

CHAPTER 4

Abacus Design

The purpose of the present chapter is to consider the design of abaci. We have seen (Chap. 2) that the abacus developed over a period of over 2000 years and that this development culminated in three types: the Russian with 10 counters per rod, all of weight 1; the Chinese with 5 unit counters and 2 counters of weight 5; and the Japanese with 4 unit counters and 1 counter of weight 5. These numbers refer to each rod. Which form of abacus should we favor?

Since the Japanese have given incomparably more attention to the development of the abacus than anyone else, and since their experts have attained the highest speed in abacus calculation, one might accept the *soroban* as the ultimate in abacus design.[1] It is possible, however, that other forms would be more congenial to European or American taste. One should also keep in mind that high-speed operation is not the sole criterion. There is also the *pedagogical* question: what is the best form of abacus to use as a child's introduction to numbers? And there is the question of abacus design for *other number bases*.

Thus it seems advisable to consider *ab initio* the whole problem of abacus design. No such general study seems to have been made previously. Various questions immediately present themselves:

1) Separate counters or a self-contained abacus?
2) Unique specification or non-unique?
3) Counters of how many values?
4) How many counters?
5) Shape of counters?
6) Size?
7) Material?
8) Allowable movement?

All eight considerations affect the design of an abacus. And after the instrument is designed and constructed, there still remains questions of operation:

9) Right-hand or left-hand operation?
10) Zero position?

It is traditional to move the counters with the right hand. For instance, the Japanese soroban is held in the left hand, and beads are manipulated with right thumb and forefinger.[1] Even when the instrument rests on a table or desk, a firm grip with the left hand is required at all times. Probably this custom originated before the advent of paper and when the results of computation were transmitted orally. With modern procedure, however, the operator reads the numbers from a sheet of paper before him and records the results on the sheet. To require the right hand to do all the work—both abacus manipulation and writing—with the left hand immobilized, is not efficiency. The remedy is to operate the abacus with the left hand, leaving the right hand free for writing. That the left hand is capable of this job is attested, for instance, by its marvellous facility in fingering the violin.

The shift to left-hand operation involves two changes:

a) The abacus is reoriented so that the rods run from left to right. This change is not absolutely necessary, though it facilitates operation.

b) The abacus is fixed in position on the desk. Perhaps the easiest way to accomplish this end is to merely increase its mass, say by making the abacus of metal.

In this and the following chapters, therefore, we shall assume left-hand operation. The abacus is placed on the operator's left, with rods running from left to right. Of course, if the reader prefers, he may use right-hand operation, with the abacus orientation shown in Chap. 2.

Question 10) should also be mentioned: the operator must decide on a convention regarding the zero position of the counters. Usually the zero position is taken on the left, though a zero on the right is possible. This choice, of course, has no effect on the design of the abacus.

4.1 General Considerations

A form of abacus, which was used in ancient times as well as in the middle ages,[2] consists in a board or table and a set of separate counters (Sections 2.1 and 2.2). Today, one can amuse himself with such an arrangement. A convenient form of counter is the *stone* of a Japanese GO set. Obviously, however, it takes much longer to pick a stone from a pile and place it in the correct position on the board than to shift a bead along a rod. The use of separate counters is cumbersome, slow, and inconvenient. We shall therefore restrict our study to self-contained abaci with counters sliding on rods.

The next question is whether the setting of the abacus should be *unique*. For instance, if 9 unit counters[3] are used on each rod, there is one and only one way of setting the abacus to represent any given number of the Base-10 system. A 1:1 relation exists between numbers and their representations on the abacus: the specification is unique. On the other hand, the Russian abacus with 10 unit counters on each rod does not give a unique specification. Evidently, the number 10 can be represented by 10 counters on the unit rod or by 1 counter on the 10-rod. Similarly, the number 110 can be set on the abacus as 1,1,0 or as 1,0,10 or as 0,10,10.

The modern Japanese abacus gives a unique specification. On a given rod, the operator moves unit beads to the center bar to represent 1, 2, 3, or 4. To represent 5, he moves the upper counter (value 5) to the center bar, all unit counters being in the zero position. For 9, all counters are shifted to the center bar. The only way to represent 10 is by one counter on the 10-rod. The Chinese abacus, on the other hand, gives by no means a unique specification. The total value of counters on a rod is 15, so there are various ways of representing the same number.

For a number system with base b, the *total value* of counters on each rod must be $(b - 1)$ if the specification is to be unique. Since this is the smallest possible number, such a device may be called a *minimal abacus*. Thus for Base-10, the minimal abacus has 9 counters per rod if only unit counters are used; but for Base-12, 11 unit counters are employed per rod for the minimal abacus. Similarly, if unit counters and 5-counters are employed, the minimal Base-10 abacus has a total *value* of 9 per rod, obtained by using 4 unit counters and 1 counter of value 5. *Any minimal abacus gives a unique specification.*

Since a 1:1 relation exists between numbers and abacus settings for a minimal abacus, such a device may be somewhat easier to read than a non-minimal abacus. On the other hand, much greater flexibility is obtained with the latter. *The minimal abacus dictates a unique procedure that must be used in adding any given pair of numbers.* The non-minimal abacus allows the operator greater freedom so that he can, if he wishes, use the shortcuts of Chap. 3. Thus the answer to our second question is not at all evident and may depend on individual preference.

ABACUS DESIGN 77

4.2 Abacus Specification

It is convenient to classify abaci with respect to the number of counters per rod. The simplest arrangement contains only counters of unit weight. If there are m unit counters per rod and no counters of higher weight, the instrument is called an $(0, m)$ abacus. If, on the other hand, there are m unit counters and n counters of higher weight on each rod, we have an (n, m) abacus. For example, the Russian abacus is said to be $(0, 10)$, but the Japanese is $(1, 4)$. This method of specification can be extended to three weights of counters, and such an instrument may be called an (p, n, m) abacus.

In all cases, an abacus has *unit* counters. Note, however, that the weights of other counters may often be chosen more-or-less arbitrarily. The physical instrument remains the same and its specification remains the same, but its operation depends on the weights chosen for its higher counters. For instance, the Chinese abacus is designated as $(2, 5)$, the 5 referring to the unit counters and the 2 to counters of weight 5. But the operator could learn a new manipulation with the higher beads weighted at 4 or 3 or even 2. Similarly, it is customary to take the zero position *down* for the unit counters and *up* for the 5-counters.[4] If the operator prefers, however, he may employ a different convention, such as the down position for both zeros. The instrument is unaltered and is still designated as $(2, 5)$, but the operation is changed.

4.3 The $(0, m)$ Abacus

We consider first the $(0, 9)$ abacus having 9 unit counters per rod and no counters of higher value. This is a minimal abacus and is the pedagogical abacus *par excellence*. It best exhibits the structure of the Base-10 system. In counting, we move counters right, one at a time on the unit rod. In this way, we reach 9. For 10, the counters on the unit rod are moved back to zero, and a single counter is moved right on the 10-rod. Evidently, any number can be represented uniquely on the $(0, 9)$ abacus. The operation of *addition* on the abacus is mechanical. Two numbers x and y, are added by setting x on the abacus and then introducing y, digit by digit, starting at the right, just as in addition by the usual pencil-and-paper method.

For the $(0, 9)$ abacus, the rules of addition are

a) If the number of available counters at zero is equal to or greater than the digit y to be added, move y counters to the right;

b) If the number of available counters at zero is less than the digit y to be added, add 10 and subtract the complement of y.

78   THE ABACUS

For example, suppose that the abacus is set at $x = 4$, and we add $y = 5$. The number of available counters is $9 - x = 5$, so 5 are moved. But if 6 were to be added instead of 5, this number would exceed the available counters. So we would move one counter on the 10-rod and would then subtract the complement of 6 by moving 4 counters on the unit rod (Fig. 4.1). This procedure is repeated for the other digits of an extended number y. An example is shown in Fig. 4.2.

The Russian abacus (0, 10) is not a minimal device, and a slight flexibility is thereby introduced into the manipulation. In adding $y = 6$ to $x = 4$, for example, the operator may either add 10 and subtract 4, as in the (0, 9), or he may merely add 6.

With the above abaci, the operator may have to shift as many as 10 beads for a single addition. This number can be reduced in either of two ways:

i) By including counters of other values.
ii) By employing complements if $y > 5$.

Let us consider ii). Suppose that $x = 9$ and that direct addition is to be used when $y = 1, \ldots 5$ but complements are to be used for higher values of y. Evidently, an $(0, m)$ abacus must have 14 beads per rod to allow such addition with $x = 9$. Here we have the apparent paradox that an *increase* in the

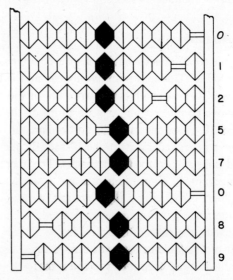

Figure 4.1   The (0, 9) abacus reading 1257089. The zero is on the left, and the fifth bead on each rod is colored to facilitate reading.

ABACUS DESIGN

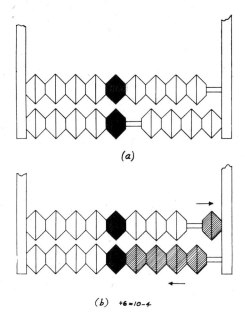

Figure 4.2 The addition $4 + 6 = 10$ on the (0, 9) abacus.

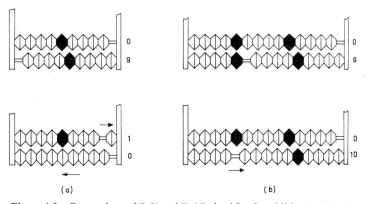

Figure 4.3 Comparison of (0, 9) and (0, 14) abaci for the addition $9 + 1 = 10$. The (0, 9) abacus requires the movement of ten beads, while the (0, 14) moves only one bead.

umber of counters allows a *decrease* in the number of counters moved. For ample, with $x = 9$ and $y = 1$ the (0, 9) abacus requires a movement of one ounter on the 10-rod and nine counters on the unit rod; but the (0, 14) quires the movement of only one counter (Fig. 4.3). A more detailed comparison is given in Table 4.1. This analysis shows that the (0, 14) abacus is superior to the (0, 9) both in the number of counters moved and in the groups of counters moved.

Table 4.1 Comparison of Two Abaci

	$x = 9$				$x = 8$			
	(0, 9)		(0, 14)		(0, 9)		(0, 14)	
y	$Wt = 10$	$Wt = 1$	$Wt = 10$	$Wt = 1$	$Wt = 10$	$Wt = 1$	$Wt = 10$	$Wt = 1$
1	1	−9	0	1	0	1	0	1
2	1	−8	0	2	1	−8	0	2
3	1	−7	0	3	1	−7	0	3
4	1	−6	0	4	1	−6	0	4
5	1	−5	0	5	1	−5	0	5
6	1	−4	1	−4	1	−4	1	−4
7	1	−3	1	−3	1	−3	1	−3
8	1	−2	1	−2	1	−2	1	−2
9	1	−1	1	−1	1	−1	1	−1
Beads moved	54		29		45		29	
Groups moved	18		13		17		13	

Note The numbers refer to bead movement on the 10-rod and on the unit rod, respectively. For instance, 1, −8 means that one counter is moved to the right on the 10-rod and 8 counters are moved to the left on the unit rod.

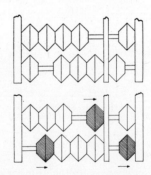

Figure 4.4 The addition $8 + 6 = 14$ on the (1, 4) abacus. The 6 can be added only by considering it as $6 = 10 − 4 = 10 − 5 + 1$, thus requiring the movement of beads in three distinct parts of the instrument. (The zero for the 5-beads is here taken on the right.)

Table 4.1 refers to $x = 9$ and $x = 8$. The same pattern holds for $9 \geq x \geq 4$: in no case are more than 5 counters moved on the (0, 14) abacus (Table 4.2). For $x < 4$, however, this simplicity can no longer be maintained. With $x = 3$ and $y = 6$, for instance, we would like to add 10 and subtract 4. But only 3 counters are available, so the operator is forced to move 6 counters.

Table 4.2   Criteria

	(0, 9)		(0, 14)	
x	Beads moved	Groups moved	Beads moved	Groups moved
0	45	9	45	9
1	38	10	38	10
2	33	11	33	11
3	30	12	30	12
4	29	13	29	13
5	30	14	29	13
6	33	15	29	13
7	38	16	29	13
8	45	17	29	13
9	54	18	29	13
Totals	375	135	320	120
$\beta =$	4.17		3.56	
$\gamma =$		1.50		1.33
$\zeta =$	0.50		0.32	
$\theta =$	0		0	

This illustrates a peculiarity of most abaci: *the procedure to be employed in the addition of y depends not only on y but also on the x already set on the instrument*. To compare abaci, therefore, one considers all values of x from 0 to 9; and for each x, he prepares a table showing what beads are moved for each value of y, where $1 \leq y \leq 9$. For each value of x, therefore, we obtain a table such as those of Table 4.1; and from it we obtain the total beads moved and the total groups moved.

This procedure is repeated for each value of x and the results are summarized as in Table 4.2. The (0, 9) abacus requires a grand total of 375 bead movements, but the (0, 14) requires only 320. A *criterion of abacus excellence* is obtained by dividing these totals by the number of additions considered

(90). This gives $\beta = 4.17$ average number of beads moved per addition for the (0, 9) and $\beta = 3.56$ for the (0, 14).

A related criterion γ evaluates the average number of groups moved per addition. Auxiliary criteria are ξ, the number of 2-move operations (expressed as a fraction of the total number of additions) and θ, the number of 3-move operations. With the (0, 9) abacus, just 50 percent of the 90 operations require movement of a bead on the 10-rod as well as movement of beads on the unit rod. Thus $\zeta = 0.50$. The corresponding criterion for the (0, 14) is $\zeta = 0.32$. With either instrument, $\theta = 0$, but we shall find that this criterion is not necessarily zero for the (n, m) abacus.

Perhaps a better criterion would be *speed of operation*. But this is a difficult thing to evaluate, particularly since it depends so much on the operator. To compare the speeds of a tyro on different instruments would be meaningless; to compare the speeds of experts is impossible at present because experts are available on only one or two kinds of abaci and it takes years of practice to reach real proficiency.[5]

4.4 The (n, m) Abacus

The most popular minimal abacus is the (1, 4) Japanese soroban mentioned in Chap. 2. Here the movement of counters is kept low by the use of counters of two values.

For any setting x of the instrument and any given y, there is just one way of making an addition. When the number of available counters on zero is equal to or less than the added digit, addition is exactly as with the (0, 9) abacus. But if this condition is not satisfied, the addition process becomes more complicated. There are six possibilities:

1. Move unit counters upward

for the additions

0 + 1	1 + 1	2 + 1	3 + 1
2	2	2	
3	3		
4			

5 + 1	6 + 1	7 + 1	8 + 1
2	2	2	
3	3		
4			

2. *Move 5-counter downward and unit counters upward*

   for the additions

0 + 5	1 + 5	2 + 5	3 + 5	4 + 5
6	6	6	6	
7	7	7		
8	8			
9				

3. *Move 5-counter downward and unit counters downward*

   for the additions

1 + 4	2 + 3	3 + 2	4 + 1
	4	3	2
		4	3
			4

4. *Move one counter upward on 10-rod, move unit counters downward on unit rod*

   for the additions

1 + 9	2 + 9	3 + 9	4 + 9
	8	8	8
		7	7
			6

6 + 9	7 + 9	8 + 9	9 + 9
	8	8	8
		7	7
			6

5. *Move one counter upward on 10-rod,*

   move 5-counter upward ⎫
   move unit counters upward ⎬ *on unit rod*

   for the additions

5 + 5	6 + 5	7 + 5	8 + 5	9 + 5
6	6	6	6	
7	7	7		
8	8			
9				

84   THE ABACUS

6. *Move one counter upward on 10-rod*,

move 5-counter upward
move unit counters down- } *on unit rod*
ward

for the additions

6 + 4	7 + 3	8 + 2	9 + 1
	4	3	2
		4	3
			4

This procedure seems complicated, but one quickly learns which of the six possibilities must be employed for any given addition. There remains, however, the unfortunate fact that in many cases counters must be moved in two or even three parts of the abacus to effect a single addition of integers. An instance is illustrated in Fig. 4.4.

Let us see if the procedure can be simplified by increasing the number of counters. The (1, 5) soroban, which was used in Japan up to about 1930,[1] is not a minimal abacus and consequently allows a degree of freedom in the movement of the counters. Of course, the (1, 5) can be used exactly like the (1, 4) by disregarding the fifth bead. But, for instance in adding $y = 1$ to $x = 4$, we can simply move a unit bead upward in the (1, 5), while the (1, 4) requires the more-complicated movement of a 5-bead and four unit beads. Analysis shows that the average number of groups of counters moved per addition is 1.78 for the (1, 5), against 1.95 for the (1, 4). As with any non-

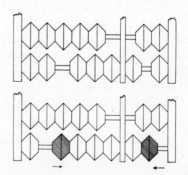

Figure 4.5 The addition $8 + 6 = 14$ on the (2, 5) abacus. The 6 is now added according to $6 = 5 + 1$, which is simpler than in Fig. 4.4 (The zero for the 5-beads is again taken on the right.)

ABACUS DESIGN 85

minimal abacus, the value 1.78 is not unique but depends somewhat on how the abacus is used.

The conventional Chinese abacus is (2, 5). It might seem that the additional 5-counter is completely superfluous. Analysis shows, however, that the extra counter greatly increases flexibility and generally *allows the elimination of 3-move additions*, which are one of the principal faults of the (1, 4) abacus. For example, with $x = 8$ on the (1, 4) abacus, the addition of $y = 6$ requires the movement of a 10-bead, a 5-bead, and a unit bead. The corresponding addition on the (2, 5) abacus requires merely the movement of a 5-counter and a unit counter (Fig. 4.5). Other additions with $x = 8$ are listed in Table 4.3. Taking all x's, we find that $\gamma = 1.56$ for the (2, 5), compared with $\gamma = 1.95$ for the (1, 4).

Table 4.3 Comparison of Three Abaci for $x = 8$

y	(1, 4)			(1, 5)			(2, 5)		
	Wt 10	5	1	Wt 10	5	1	Wt 10	5	1
1	0	0	1	0	0	1	0	0	1
2	1	−1	−3	0	0	2	0	0	2
3	1	−1	−2	1	−1	−2	0	1	−2
4	1	−1	−1	1	−1	−1	0	1	−1
5	1	−1	0	1	−1	0	0	1	0
6	1	−1	+1	1	−1	+1	0	1	1
7	1	0	−3	1	0	−3	0	1	2
8	1	0	−2	1	0	−2	1	0	−2
9	1	0	−1	1	0	−1	1	0	−1
Beads moved	27			24			19		
Groups moved	21			19			15		

Note The numbers refer to the movement of unit beads on the 10-rod, and 5-counters and unit counters on the unit rod. For instance, 1, −1, −3 means that a unit counter is moved to the right on the 10-rod, a 5-counter is moved to the left on the unit rod, and three unit counters are moved to the left on the unit rod.

Table 4.4 gives the complete summary for the three abaci. Note that the (2, 5) abacus is definitely superior to the (1, 4), both as regards the number of moved beads and the number of moved groups. Also, the troublesome 3-move operations of the (1, 4) are completely eliminated in the (2, 5).

Table 4.4 Criteria for (n, m) Abaci

x	(1, 4)		(1, 5)		(2, 5)	
	Beads moved	Groups moved	Beads moved	Groups moved	Beads moved	Groups moved
0	25	13	25	13	25	13
1	20	14	22	11	22	11
2	19	15	19	14	19	14
3	22	16	20	15	20	15
4	29	17	25	16	25	16
5	30	18	33	17	25	13
6	25	19	26	17	25	13
7	24	20	23	18	20	14
8	27	21	24	19	19	15
9	34	22	29	20	22	16
Totals	255	175	246	160	222	140
$\beta =$	2.84		2.74		2.47	
$\gamma =$		1.95		1.78		1.56
$\zeta =$	0.51		0.44		0.48	
$\theta =$	0.22		0.18		0	

4.5 The (2, 2, 4) Abacus

There is no reason why the value of the counters should be fixed at 1 and 5. For instance, we could consider an abacus with 4 unit beads per rod, 2 beads of weight 2, and 2 beads of weight 5. This (2, 2, 4) abacus is shown in Fig. 4.6. It is convenient to take the zero position to the left in all cases, so a single tilt of the instrument sets all counters at zero. The addition of 8 + 2 requires the movement of only one counter. With the (1,4) abacus, this addition requires movements in three separate parts of the instrument: the movement of a unit bead on the 10-rod, the movement of a 5-counter, and the movement of 3 unit counters on the unit rod. For all possible integers, the average number of counters moved per addition is $\beta = 1.72$ on the (2, 2, 4), compared with $\beta = 2.84$ on the (1,4) abacus.

4.6 Negative Counters

Another idea is to weight some of the counters negatively. Such a modification is relatively sophisticated: it would not have been made in antiquity. It fits rather nicely, however, with the idea of complements. Suppose that we formulate the rule that addition of integers from 1 to 5, inclusive, shall be made directly but that the addition of 6 to 9 shall always be made by using complements. None of the previously mentioned abaci allow such a rule for all values of x. But if we include a set of negative counters, we can employ this rule in all cases, which should keep bead-movement low.

Figure 4.6   The (2, 2, 4) abacus, with beads of weight 1, 2, and 5. The zero position is taken on the left in all cases.

A possible construction utilizes the (0,14) abacus of Section 4.3 but adds a section to the right of the other. This section contains, per rod, four beads of weight -1. Let us take the zeros of the positive counters to the left and the zero of the negative counters to the right. Then *addition* is always effected by moving a bead to the right, *subtraction* by moving a bead to the left.

THE ABACUS

Suppose that the abacus is set at zero and we wish to add 5. This is accomplished by moving 5 positive beads to the right. To add 6, however, we use the complement, moving 1 positive bead on the 10-rod and 4 negative beads on the unit rod. Similarly, 9 is added by adding 1 on the 10-rod and -1 on the unit rod. In this case, only two counters are moved, as contrasted with the movement of 9 counters on the (0, 9) abacus. The average counters moved per addition is 3.22 for the $(-4, 0, 14)$ instrument and 4.17 for the (0, 9).

To reduce the rather excessive total number of beads, we can introduce two values, as in the $(-4, 2, 4)$ abacus of Fig. 4.7. Here all positive beads have zero to the left, and negative beads have zero to the right. As before, this means that moving right always adds, moving left subtracts.

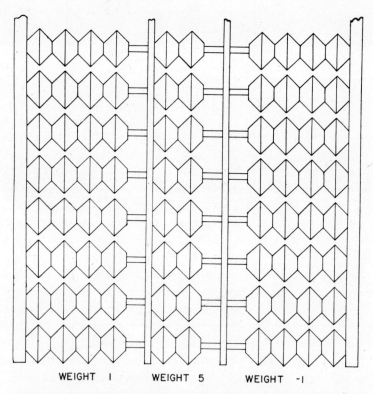

Figure 4.7 The $(-4, 2, 4)$ abacus, with beads of weight 1, 5, and -1. The positive beads have zero positions on the left. The negative beads have zeros on the right so that movement of any bead to the right always *increases* the number set on the instrument.

Operation is very similar to the $(-4,0,14)$: we can always add $1, \ldots 5$ directly and $6, \ldots 9$ by complement. Results are essentially as with the $(-4,0,14)$, the average number of counters moved per addition being 2.78. A general compilation of results is given in Table 4.5.

Table 4.5 General Comparison of Abaci for Base-10

Abacus	β	γ	ζ	θ
(0, 9)	4.17	1.39	0.50	0
(0, 10)	4.07	1.40	0.41	0
(0, 14)	3.56	1.33	0.32	0
(1, 4)	2.84	1.95	0.51	0.22
(1, 5)	2.74	1.78	0.44	0.18
(2, 5)	2.47	1.56	0.58	0
(2, 9)	2.75	1.42	0.44	0
(2, 2, 4)	1.72	1.64	0.63	0
(−4, 0, 14)	3.22	1.45	0.44	0
(−4, 2, 4)	2.78	1.67	0.67	0

Note For non-minimal abaci, the numerical values depend slightly on the choice of operating procedure.

On the basis of this table, preferable forms of abaci seem to be the (0,9), the (2,5), the (2,2,4), and the (−4,2,4). The (0,9) abacus is most intimately connected with the Base-10 number system and is simplest to learn. The (2,2,4) is less straightforward, since it is associated also with Base-2 and Base-5, but it has the smallest amount of bead movement. The (−4,2,4) allows full use of complements, so that a definite program can be maintained, irrespective of the original number x set on the abacus.

Note that none of the traditional forms of abacus (the Japanese, the Chinese, the Russian) appears very favorably, according to this analysis. The soroban, with its high bead movement and its triple moves, seems particularly reprehensible. But this is the instrument that beat the modern desk calculator! Of course, it is very difficult to correlate *speed* with anything in Table 4.5. Perhaps an operator with sufficient experience on one of the other abaci could beat the best soroban operators!

4.7 Other Considerations

Of the list of design considerations, given at the beginning of the chapter, we have now handled the first four items. There remain the shape, size, material, and allowable movement of the counters. Regarding *shape*, the Chinese abacus employs a bead that is approximately an oblate spheroid. An improved form with sharp edge was developed in Japan.[1] Figure 4.8 shows

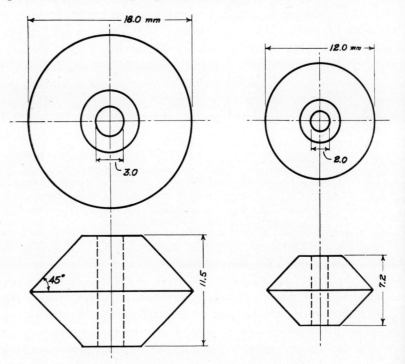

Figure 4.8 Typical dimensions of abacus beads.

two sizes, the smaller being the more common. The traditional material is wood, though metal, glass, and plastic are possibilities. Allowable movement is approximately 50 mm in the Chinese abacus. As the operator becomes more proficient, however, he prefers a smaller movement. A value of 6 mm is common in the Japanese instruments.*

---

* For the large beads of Fig. 4.8, unit beads have 8 mm movement and 5-beads have 7 mm movement. For the small beads, movements are 6 mm and 4 mm respectively.

4.8 Number Bases[6]

All of the abaci listed in Table 4.5 are for the Base-10 system. But the designs can be changed in a very simple and obvious manner to apply to any other number base. Here is a fascinating potential application of the abacus, allowing us to work in other number systems with almost the facility to which we are accustomed in Base-10. For instance, a simple way of familiarizing oneself with Base-7 is to use a simple abacus like (0,9) but with 6 beads per rod instead of 9. Or if one needs the duodecimal system, he can arrange an abacus with 11 beads per rod instead of 9. For larger bases, such as 16, 24, 30, or 60, this simple type of abacus is hardly practicable. But with 60, for example, one can use a (5,9) abacus, where the familiar 10-system is used as a sub-base. In this case, the 9 refers to unit beads and the 5 to beads of weight 10. Or other analogs of Table 4.5 are possible.

Even with Base-12, there is no need to restrict oneself to the (0,11) abacus: for instance, (2,6), (4,4), or (2,2,5) could be used. Here the last number in each parentheses refers to the number of unit beads per rod, the 2 in (2,6) gives the number of counters of weight 6, the first 4 in (4,4) specifies the number of weight 4, and (2,2,5) refers to weights 6, 3, and 1, respectively. This subject will be amplified in later chapters.

4.9 General Theory, Minimal Abaci

We now consider the general theory of the abacus, applicable to any base. For a *minimal abacus*, the total value of all beads on the unit rod is $(b - 1)$, where b is the number base. If each bead has unit value or weight, we have an abacus designated as

$$(0, m)$$

where m is the number of beads per rod. Evidently, for any minimal abacus of this type, the number of beads per rod is

$$m = b - 1. \tag{4.1}$$

Analysis shows that increasing the number of beads beyond this value has very little beneficial effect and merely results in a clumsy abacus. Thus we recommend that abaci of type $(0,m)$ be of the minimal kind with beads per rod equal to one less than the base. For instance,

b	m	
6	5	beads per rod
8	7	
10	9	
12	11	

It is doubtful if the $(0,m)$ abacus should be used for bases greater than 12 because other arrangements allow more compact instruments.

For the larger bases, an (n,m) abacus may be preferable. Here m refers to the number of unit beads per rod, as previously, and n refers to the number of beads of weight w:

$$\text{Wt} = w \quad \text{Wt} = 1$$
$$\diagdown \quad \diagup$$
$$(n, m)$$

Such an abacus utilizes the base b and a sub-base b' equal to the weight w. It is convenient to introduce the ratio ξ of the two bases:

$$\xi = b/b' \tag{4.2}$$

and to choose b' such that ξ is an integer. For a minimal abacus, the total weight per rod is still $(b - 1)$, so

$$b - 1 = nb' + m. \tag{4.3}$$

Also, n cannot be greater than $(\xi - 1)$ if Eq. (4.3) is to be satisfied:

$$n = \xi - 1. \tag{4.4}$$

Therefore,

$$b - 1 = \frac{b}{\xi}(\xi - 1) + m$$

or

$$m = \frac{b}{\xi} - 1 = b' - 1. \tag{4.5}$$

For example, if the sub-base is one-half of base b and if $b = 10$, then

$$\xi = 2, \quad n = 1, \quad m = 4,$$

ABACUS DESIGN

Table 4.6 Minimal Abaci of Type (n, m)

b	$b' = w$	ξ	n	M	N
10	5	2	1	4	5
12	6	2	1	5	6
	4	3	2	3	5
16	8	2	1	7	8
	4	4	3	3	6
24	12	2	1	11	12
	8	3	2	7	9
	6	4	3	5	8
30	15	2	1	14	15
	10	3	2	9	11
	6	5	4	5	9
60	30	2	1	29	30
	10	6	5	9	14

which gives the familiar Japanese soroban. Some other minimal abaci are indicated in Table 4.6.

The total number of beads per rod is

$$N = m + n = \frac{b}{\xi} + \xi - 2, \tag{4.6}$$

which is a function of the ratio ξ. The most compact abacus for a given base is obtained by minimizing N:

$$\frac{\partial N}{\partial \xi} = 0 = 1 - b/\xi_0^2$$

or

$$\xi_0 = (b)^{1/2}. \tag{4.7}$$

Thus for $b = 16$, the most compact minimal abacus is obtained with $\xi = \sqrt{16} = 4$ and $b' = w = 4$. This gives a total of $N = 6$ beads per rod; while for $\xi = 2$, we have $N = 8$.

4.10 General Theory, Non-Minimal Abaci

The great defect of the minimal (n, m) abacus is that three distinct bead movements are sometimes required to obtain the sum of two digits (Section 4.4). This defect can be eliminated by using more than the minimal number of beads. Let us consider the subject in detail.

One must remember that the abacus operation

$$S = x + y$$

depends quite as much on the number x, which is originally on the abacus, as on the number y that is to be added. For the $(1, 4)$ abacus with $b = 10$ and $x = 5$, for example, the 5-bead has been moved and is no longer available for the adding of y. Suppose that $y = 8$. We cannot add 8 beads because only 4 are available. Also, we cannot move a bead on the 10-rod and subtract 2 on the unit rod because all unit beads are at zero. The only possibility, therefore, is to move a bead on the 10-rod, move the 5-bead to zero on the unit-rod, and then add 3 unit beads:

$$+8 = 10 - 5 + 3.$$

It is this kind of triple move that we are trying to eliminate.

Evidently, we need only keep a 5-bead in reserve to remedy the difficulty. In other words, let us increase n to 2 for this $(1, 4)$ abacus. In general, n must be large enough to allow $(\xi - 1)$ beads for x, plus $(\xi - 1)$ beads for y, or

$$n = 2(\xi - 1). \tag{4.8}$$

Equation (4.8) gives the *number of beads of weight w* required for any base b.

How many *unit beads* should be used per rod? With $x = 0$ and y having its maximum value $(b - 1)$, we use $(n - 1)b'$-beads. The remainder of y must be taken care of by m unit beads, so

$$m = b' - 1.$$

Detailed analysis shows that this number of beads is sufficient also for other values of x and does not lead to triple moves. However, a somewhat more flexible abacus is obtained with an extra unit bead. Let us therefore standardize on

$$m = b'. \tag{4.9}$$

ABACUS DESIGN

The total number of beads per rod is then

$$N = m + n = 2\xi + b/\xi - 2. \quad (4.10)$$

For a minimum number of beads,

$$\frac{\partial N}{\partial \xi} = 0 = 2 - b/\xi_0^2$$

or

$$\xi_0 = (b/2)^{1/2}. \quad (4.11)$$

Some non-minimal (n,m) abaci that eliminate all triple-moves are listed in Table 4.7.

Table 4.7 Non-minimal Abaci of Type (n, m)

| Base | Weight | ξ | Number of Beads per Rod | | |
b	$w = b'$		n	m	N
6	3	2	2	3	5
8	4	2	2	4	6
10	5	2	2	5	7
12	6	2	2	6	8
	4	3	4	4	8
16	8	2	2	8	10
	4	4	6	4	10
24	12	2	2	12	14
	8	3	4	8	12
	6	4	6	6	12
30	15	2	2	15	17
	10	3	4	10	14
	6	5	8	6	14
60	30	2	2	30	32
	15	4	6	15	21
	12	5	8	12	20
	10	6	10	10	20

For the larger bases, there may be advantages in using two sub-bases instead of one:

$$\text{Wt} = b'$$
$$\text{Wt} = b''$$
$$\text{Wt} = 1 \quad (p, n, m).$$

The principal base remains b, but we now have sub-bases b' and b''. We also introduce the ratios

$$\xi = b/b', \quad \xi' = b'/b''. \qquad (4.12)$$

Using the same argument that was employed with the (n,m) abacus, we write

$$p = 2(\xi - 1), \qquad (4.13)$$

$$n = 2(\xi' - 1). \qquad (4.14)$$

Analysis shows that triple moves are eliminated if

$$m = b' - 1. \qquad (4.5)$$

The total number of beads per rod is

$$N = m + n + p = \frac{b}{\xi} + 2(\xi + \xi') - 5. \qquad (4.15)$$

If we fix ξ', then the minimum number of beads is obtained by the same relation, Eq. (4.11) that was used with the (n,m) abacus. Data are given in Table 4.8.

Table 4.8 The (p, n, m) Abacus

b	b'	b''	ξ	ξ'	p	n	m	N
10	5	2	2	2.5	2	3*	4	9
12	6	3	2	2	2	2	5	9
16	8	4	2	2	2	2	7	11
24	12	6	2	2	2	2	11	15
	8	4	3	2	4	2	7	13
30	15	5	2	3	2	4	14	20
	10	5	3	2	4	2	9	15
60	30	10	2	3	2	4	29	35
	10	5	6	2	10	2	9	21

* Satisfactory with $n = 2$.

Abaci can be extended to four bases if desired. The designation then becomes

Wt = b'
Wt = b''
Wt = b''' Wt = 1 (q, p, n, m).

ABACUS DESIGN

The four bases are $b, b', b'', $ and b''', and the ratios are

$$\xi = b/b', \quad \xi' = b'/b'', \quad \xi'' = b''/b'''. \tag{4.16}$$

By similarity with the preceding equations, we have

$$\left.\begin{aligned} q &= 2(\xi - 1), \\ p &= 2(\xi' - 1), \\ n &= 2(\xi'' - 1), \\ m &= b' - 1. \end{aligned}\right\} \tag{4.17}$$

Also,

$$N = m + n + p + q = b' + 2(\xi + \xi' + \xi'') - 7. \tag{4.18}$$

Some possibilities are listed in Table 4.9.

Table 4.9 The (q, p, n, m) Abacus

b	b'	b''	b'''	ξ	ξ'	ξ''	q	p	n	m	N
12	6	3	2	2	1.5	2	2	1	5	10
16	8	4	2	2	2	2	2	2	2	7	13
24	12	6	3	2	2	2	2	2	2	11	17
24	8	4	2	3	2	2	4	2	2	7	15
30	10	5	2	3	2	2.5	4	2	3	9	18
60	10	5	2	6	2	2.5	10	2	3	9	24

Finally, we consider the use of beads with negative values, giving the abacus

$$\begin{aligned} \text{Wt} &= -1 \\ \text{Wt} &= b' \\ \text{Wt} &= +1 \end{aligned} \quad (-m, n, m).$$

This merely amounts to making an (n, m) abacus with an extra set of m beads. It is found, however, that the extra unit bead that was convenient in the (n, m) abacus is now redundant. Therefore,

$$n = 2(\xi - 1), \tag{4.8}$$

$$m = b' - 1. \tag{4.5}$$

Moon (0196)

98   THE ABACUS

The total number of beads per rod is now
$$N = 2m + n = 2(b/\xi + \xi) - 4. \tag{4.19}$$
Also,
$$\frac{\partial N}{\partial \xi} = 0 = 1 - b/\xi_0^2$$
or
$$\xi_0 = (b)^{1/2}. \tag{4.7}$$

Some designs are listed in Table 4.10.

Table 4.10 Non-Minimal Abaci of Type $(-m, n, m)$

Base b	Weight $w = b'$	ξ	Number of Beads per Rod		
			n	m	N
10	5	2	2	4	10
12	6	2	2	5	12
	4	3	4	3	10
16	8	2	2	7	16
	4	4	6	3	12
24	12	2	2	11	24
	8	3	4	7	18
	6	4	6	5	16
30	15	2	2	14	30
	10	3	4	9	22
	6	5	8	5	18
60	15	4	6	14	30
	10	6	10	9	28

The tables have been calculated for representative number bases. Obviously, abaci can be made equally well for other bases, the number of beads being easily calculated. A summary of the general equations is as follows

Minimal abaci

Type $(0, m)$. Base b.
$$m = b - 1. \tag{4.1}$$
Type (n, m). Bases b, b'.
$$n = \xi - 1, \tag{4.4}$$
$$m = b' - 1, \tag{4.5}$$

where $\xi = b/b'$. Also,
$$N = b' + \xi - 2, \tag{4.6}$$
$$\xi_0 = (b)^{1/2}. \tag{4.7}$$

Non-minimal abaci

Type (n, m). Bases b, b'.
$$n = 2(\xi - 1), \tag{4.8}$$
$$m = b', \tag{4.9}$$
$$N = 2\xi + b' - 2, \tag{4.10}$$
$$\xi_0 = (b/2)^{1/2}. \tag{4.11}$$

Type (p, n, m). Bases b, b', b''.
$$p = 2(\xi - 1), \tag{4.13}$$
$$n = 2(\xi' - 1), \tag{4.14}$$
$$m = b' - 1, \tag{4.5}$$
$$N = b' + 2(\xi + \xi') - 5. \tag{4.15}$$

Type (q, p, n, m). Bases b, b', b'', b'''.
$$\begin{aligned} q &= 2(\xi - 1), \\ p &= 2(\xi' - 1), \\ n &= 2(\xi'' - 1), \\ m &= b' - 1, \\ N &= b' + 2(\xi + \xi' + \xi'') - 7. \end{aligned} \tag{4.17}$$
$$\tag{4.18}$$

Type $(-m, n, m)$. Bases b, b'.
$$n = 2(\xi - 1), \tag{4.8}$$
$$m = b' - 1, \tag{4.5}$$
$$N = 2(\xi + b') - 4, \tag{4.19}$$
$$\xi_0 = (b)^{1/2}. \tag{4.7}$$

The choice of an abacus for a given base is largely a personal matter. Some recommended designs, however, are given in Table 4.11.

Table 4.11 Recommended Abaci

Bases			Beads per rod			
b	b'	b''	p	n	m	N
6	–	–	0	0	5	5
8	–	–	0	0	7	7
10	–	–	0	2	5	7
12	6	–	0	2	6	8
12	6	3	2	2	5	9
16	8	–	0	2	8	10
24	8	–	0	4	8	12
30	10	–	0	4	10	14
60	10	–	0	10	10	20
60	10	5	10	2	9	21

References

1. Japan Chamber of Commerce and Industry, *Soroban, the Japanese abacus: its use and practice*, Charles E. Tuttle Co., Rutland, Vt., (1967).
2. D. E. SMITH, *History of mathematics*, chap. 3, vol. 2, Ginn and Co., Boston, (1925);
   F. P. BARNARD, *The casting-counter and the counting board*, Oxford Univ. Press, (1916);
   K. MENNINGER, *Zahlwort und Ziffer*, F. Hirt, Breslau, (1934);
   O. ORE, *Number theory and its history*, McGraw-Hill Book Co., New York, (1948).
3. The term *unit counter* refers to the elementary counter or bead, irrespective of what rod it is on. If a unit counter is on the unit rod, its value is 1; if on the ten rod, its value is 10, etc. Other counters have *weight* or *value* of w times that of the unit counter on the same rod. The usual weight of a non-unit counter is 5, though other values may be assigned if desired.
4. Since the abacus is usually operated with the plane of the rods in a horizontal position, "up" and "down" are metaphorical. By "up" we mean of course "farther from the operator" and by "down" "nearer the operator".
5. T. KOJIMA, *The Japanese abacus, its use and theory*, Charles E. Tuttle Co., Rutland, Vt., (1957);
   Y. YOSHINO, *The Japanese abacus explained*, Dover Publications, New York, (1963).
6. J. LESLIE, *The philosophy of arithmetic*, Edinburgh, (1820);
   O. ORE, *Number theory and its history*, McGraw-Hill Book Co., New York, (1948);
   F. E. ANDREWS, *New numbers*, Harcourt, Brace, and World, New York, (1935);
   G. S. TERRY, *Duodecimal arithmetic*, Longmans, Green, and Co., London, (1938);
   J. ESSIG, *Douze notre dix futur*, Dunod, Paris, (1955);
   R. C. WEAST and S. M. SEEBY, *CRC handbook of tables for mathematics*, Chem. Rubber Co., Cleveland, (1967).

CHAPTER 5

The (0, *m*) Abacus

The remainder of the book is devoted to detailed instructions for the operation of the most promising forms of abacus. Each chapter is largely independent of the others, so the reader can turn immediately to the particular abacus that interests him and can skip the chapters that deal with other abaci.

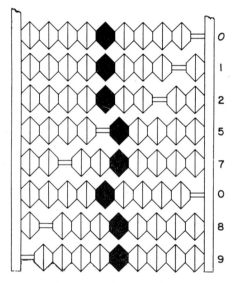

Figure 5.1   The (0, 9) abacus.

Let us first consider the (0,9) abacus used for Base-10 calculations. An instrument of this kind is shown in Fig. 5.1. It consists of a frame with a number of parallel rods, each having 9 unit counters. The abacus is placed on the desk, to the left of the operator. The rods run from left (zero) to right. Beads are operated with the index finger of the left hand. Since the zero position of the counters is taken to the left, the abacus is set at zero by momentarily raising the right side so that beads slide against their left stop.

The (0,9) abacus has pedagogical value: its construction is more closely related to the structure of the decimal system than is any other abacus. Not only do we have a rod for each place in the positional number system, but we have nine beads on each rod, corresponding to the digits 1, 2, ... 9.

5.1 Addition

The number x is set on the abacus. Another number y is to be added to x. This operation is performed in a mechanical way, without thought and without a memorized addition table.

An example is indicated in Fig. 5.2. Here $x = 86193$ and $y = 12704$. The number x is first set on the abacus as shown in Fig. 5.2a. In (b), 4 counters

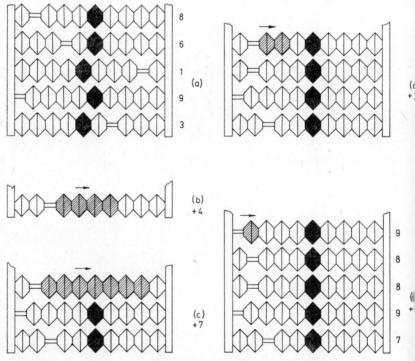

Figure 5.2 The sum 86193 + 12704 = 98897 on the (0, 9) abacus. The number 86193 is set on the abacus in (a). The digits of 12704 are then applied one by one, and the sum is read in (e).

THE (0, *m*) ABACUS 103

are moved to the right on the lowest rod. In (c), 7 counters are moved on the third rod. This process is continued until (f) is reached, when the sum 98897 is read from the abacus. As previously stated, the operation is purely mechanical: the operator does not think in (b) "four plus three is seven"— he merely moves four beads.

Addition is *commutative*, so exactly the same result is obtained whether we add $x + y$ or $y + x$. Either number can be set on the abacus and the other added to it. In practice, however, there is often an advantage in doing the addition in one way rather than the other. If one of the numbers has less digits than the other, this smaller number is taken as y. Similarly, if one number contains zeros or ones, it may be taken advantageously as y rather than x.

The above example is particularly simple because the sum of digits on each rod is less than 10. A more realistic case is shown in Fig. 5.3, where 47638 is added to 83217. The latter number is set on the abacus in (a). On the bottom rod, we next add 8. But since 8 available counters in the zero position are

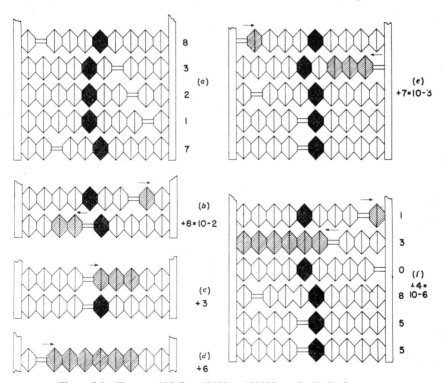

Figure 5.3 The sum 83217 + 47638 = 130855 on the (0, 9) abacus.

not available on this rod, we think $8 = 10 - 2$ and move one counter on the second rod and two counters (to the left) on the first rod (Fig. 5.3b). A similar condition occurs in (e), where 7 is added by adding 10 and subtracting 3. The final result is shown in (f), the sum being 130855.

These examples indicate the extreme simplicity of addition on the (0,9) abacus. Since this is a minimal instrument, there is a 1:1 correspondence between numbers and counter positions. Also, for the addition of any two digits, the procedure is unique. As noted in Section 3.2, the operator has no choice:

a) If the number of available counters at zero is equal to or greater than the digit y to be added, move this number of counters to the right,

b) If the number of available counters at zero is less than the digit y to be added, add 10 and subtract the complement of y.

Before going further, the reader is urged to practice addition on the (0,9) abacus until he is confident of getting the correct answer. Suggested problems are given in Appendix A. Answers are listed in Appendix B.

5.2 Subtraction

Since subtraction is merely negative addition, the procedure is obvious. Consider Fig. 5.4, for instance. Here 47638 is subtracted from 83217. The latter number is set on the abacus in (a). On the bottom rod, 8 is to be subtracted: 8 beads are to be moved to the left. But only 7 beads are available, so the complement of 8 must be added: $-8 = -10 + 2$. As shown in Fig. 5.4b, one bead is moved to the left (subtracted) on the 10-rod, and 2 beads are moved to the right (added) on the unit rod. The procedure is continued until at (f) the difference 35579 is read from the abacus. For each pair of digits, x and y, the abacus operation is determined uniquely, depending on the number of available beads and on the digit to be subtracted. Again the reader is urged to practice until the manipulation of the beads becomes automatic. Addition and subtraction are the basic operations with the abacus; and until they are mastered, it is inadvisable to proceed to multiplication or division.

5.3 Multiplication

Users of mechanical desk calculators (Chap. 1) know that these devices are not multiplying machines or even adding machines.[1] Basically, they are *counting machines*. They do not even know the addition table: they merely

THE $(0, m)$ ABACUS

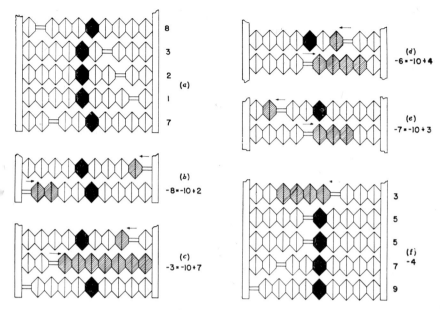

Figure 5.4 The difference $83217 - 47638 = 35579$ on the $(0, 9)$ abacus.

count 1, 2, 3, ..., though they may do this counting simultaneously on 8 or 10 dials. The abacus, likewise, is fundamentally a counting machine. In both cases, *addition* is performed mechanically by counting.

Multiplication on a desk calculator can be performed only by repeated additions. Thus if the operator wants to multiply by 9, he holds down the plus bar while the motor drives the machine through 9 complete additions of the number that is set on the keyboard. Even the so-called "automatic multiplication", obtainable with the most expensive models, is only repeated addition.

In other words, *the ordinary desk calculator does not know the multiplication table* and there is no way to teach it. If he wished to do so, the abacus operator could also multiply by repeated addition. But, fortunately, he has a brain and he knows the multiplication table. So multiplication on the abacus is performed primarily in the brain of the operator, the beads being employed merely to add the partial products.

Consider the multiplication of a number x (the multiplicand) by y (the multiplier). Let s be the number of digits in x and t the number of digits in y. Using the memorized multiplication table, one takes all possible products of the digits of x by the digits of y. There are st of these partial products, which are then added to give the product xy.

For example, to multiply 46×23 we may write

$$\begin{array}{r} 46 \\ \times\,23 \\ \hline 6 \times 3 = 18 \\ 4 \times 3 = 12 \\ 6 \times 2 = 12 \\ 4 \times 2 = 8 \\ \hline 1058. \end{array}$$

The product of the right-hand digits is $6 \times 3 = 18$, which is written directly below 23. There are two partial products where one digit is in the unit column and one in the 10-column ($3 \times 4 = 12$ and $2 \times 6 = 12$). These products are written one place to the left. And there is one partial product ($2 \times 4 = 8$), where both digits are in the 10-column. This product is written two places to the left. Since $s = t = 2$, there are $st = 4$ partial products.

This example illustrates the *basic method* of abacus multiplication. Each partial product of integers is obtained mentally. But instead of writing it, as in the above example, we set it on the abacus. As soon as the st partial products have been set on the abacus, we can read from it the final answer. Thus both the writing of the partial products and their *mental* addition are eliminated when the abacus is used.

Four methods of abacus multiplication will be considered:

1. Repeated addition on the abacus, using the procedure of Section 5.1.
2. Basic method, as outlined above, with st partial products.
3. Doubling and halving (Section 3.7).
4. Napier's rods.

Method 1. is that of the desk calculator. For example, 46×23 would be written with three 46's on the right and two 46's shifted one place:

$$\begin{array}{r} 46 \\ \times\,23 \\ \hline 46 \\ 46 \\ 46 \\ 46 \\ 46 \\ \hline 1058 \end{array}$$

THE (0, m) ABACUS

The method reduces all multiplication to simple abacus addition. No multiplication table is employed and the operator need do no thinking. The method is not recommended, however, because of the large number of items that must be added.

In the *basic method* 2., partial products are obtained mentally and are set on the abacus, proceeding from bottom to top. Because of the close relation between written multiplication and abacus multiplication, we shall preface each example by showing how it would be done by the familiar pencil-and-paper method, followed by a description of abacus procedure. The reader understands, of course, that in multiplying with the abacus *we do not write the partial products but only the final results.*

As a simple example of Method 2., take 46×23 on the (0,9) abacus. The operator thinks $6 \times 3 = 18$ and sets this number on the abacus, Fig. 5.5a. He next thinks $4 \times 3 = 12$ and adds this number, shifted one place to the left, Fig. 5.5b. This finishes with the 3 of the multiplier. Changing to the 2 of

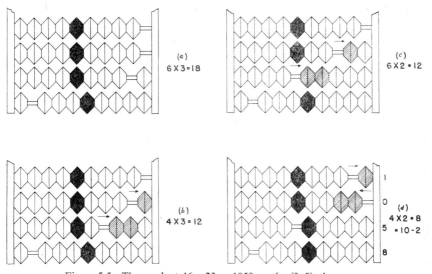

Figure 5.5 The product $46 \times 23 = 1058$ on the (0, 9) abacus.

the multiplier, the operator thinks $6 \times 2 = 12$ and moves the necessary counters (c). Finally, he takes $4 \times 2 = 8$ and incorporates this into the abacus reading. The product 1058 is then read directly from the instrument (d).

As another example, consider the product

$$1936 \times 874.$$

In terms of elementary partial products, the multiplication could be written

$$\begin{array}{r} 1936 \\ 874 \\ \hline 24 \\ 12 \\ 36 \\ 4 \\ \hline 42 \\ 21 \\ 63 \\ 7 \\ \hline 48 \\ 24 \\ 72 \\ 8 \\ \hline 1692064 \end{array}$$

The reader may try this example on his (9,0) abacus.

Thus if straightforward abacus multiplication is employed for this example, *twelve* partial products must be applied to the instrument. This takes time, and there is always the possibility of adding them in the wrong column. It is not surprising, therefore, that the abacus has difficulty in competing with the desk calculator when large numbers are multiplied, though it has no difficulty in beating the desk calculator in addition.[2] After the multiplicand 1936 is set on the keyboard of the electrically driven machine, only *three* operations are necessary: the successive multiplications by the three digits 874 of the multiplier. Though the individual multiplications are ludicrously clumsy on the desk calculator, they may take less time than the twelve operations on the abacus. And of course this difference is amplified with still larger numbers.

Practice in abacus multiplication is suggested, using problems given in Appendix A or taking examples from the Japanese books.[3]

5.4 Double and Halve

The principal weakness of the basic method of abacus multiplication is associated with the large number of partial products. Methods 3. and 4. of Section 5.3 are attempts to reduce this number *st*. Of course, in the usual pencil-and-paper multiplication, we do reduce the number of partial products from st to t by writing products for the *complete multiplicand* instead of for each digit separately. As an example, we customarily write

$$\begin{array}{r} 46 \\ \times\,23 \\ \hline 138 \\ 92 \\ \hline 1058 \end{array}$$

instead of using the four products of Section 5.3. The same condensation can be effected on the abacus, provided that the operator is able to mentally evaluate the product of any number by any digit.

In the above example, he knows that $46 \times 3 = 138$, so he sets this number on the abacus. He also knows that $46 \times 2 = 92$, which he applies to the instrument. The product 1058 is then read immediately.

The foregoing illustration shows that it is not always necessary to take a partial product for each pair of digits: sometimes several digits can be considered together. For instance, take

$$94602 \times 25501.$$

Here we can write

$$\begin{array}{r} 94602 \\ \times\,25501 \\ \hline 94602 \\ 4730100 \\ 2365050 \\ \hline 2412445602 \end{array}$$

where the multiplier has been considered as made up of $25 = 100/4$, $50 = 100/2$, and 1:

$$25501 = (\tfrac{100}{4})\,1000 + (\tfrac{100}{2})\,10 + 1.$$

In this example, $st = 25$, yet we have needed only 3 partial products.

Method 3. tends to decrease the number of partial products by reducing all multiplication to doubling and halving (Section 3.7). The scheme is as follows:

$$\times 2, \text{ direct multiplication}$$
$$3 = 2 + 1$$
$$4 = 5 - 1$$
$$5 = 10/2$$
$$6 = 5 + 1$$
$$7 = 5 + 2$$
$$8 = 10 - 2$$
$$9 = 10 - 1.$$

Since the entire multiplicand can be doubled or halved mentally, the total number of operations on the abacus is never more than $2t$ and all the partial products are obtained easily.

Let us consider some of the previous examples of multiplication, using this method. For the product

$$46 \times 23,$$

we need know only that $2 \times 46 = 92$. Thus

$$46 \times 3 = 46(2 + 1) = 92 + 46,$$

and

$$\begin{array}{r} 46 \\ 23 \\ \hline 92 \\ 46 \\ 92 \\ \hline 1058. \end{array}$$

The operation on the abacus is obvious.

For the product

$$1936 \times 874,$$

we need only combinations of

$$1936,$$
$$1936 \times 2 = 3872,$$
$$19360/2 = 9680.$$

Thus
$$1936 \times 4 = 1936(5) - 1936$$
$$= 9680 - 1936,$$
$$1936 \times 7 = 1936(5) + 1936(2)$$
$$= 9680 + 3872,$$
$$1936 \times 8 = 19360 - 1936(2)$$
$$= 19360 - 3872.$$

The additions and subtractions are done on the abacus. The complete product may be written

$$\begin{array}{r} 1936 \\ \times\ 874 \\ \hline 9680 \\ -1936 \\ 9680 \\ +3872 \\ 19360 \\ -3872 \\ \hline 1692064. \end{array}$$

Thus the original twelve partial products are reduced to six in this modified abacus multiplication.

5.5 Napier's Rods

We have seen that the ideal shortcut in abacus multiplication would reduce the st partial products of Method 2. to t partial products. This radical improvement is not obtained by Method 3., which may require $2t$ products. It is conceivable that the operator can train himself to obtain mentally the product of any number by any digit,[4] thus determining the required t numbers. Such a training, however, is far from easy, even in Base-10. And when we introduce other bases, the problem becomes even worse. The best solution of a general nature appears to be the introduction of Napier's rods (Section 3.11).

A set of rods is easily made for any specific base b. The t partial products are then read directly from the rods,[5] and the abacus is employed to add these products.[6] Napier's rods for Base-10 are shown in Fig. 3.3.

Let us consider the familiar example, 1936×874. Rods are arranged as shown in Fig. 5.6, from which we read the partial products 7744, 13552, and 15488. The complete product then requires only these three and may be written

$$\begin{array}{r} 1936 \\ \times\ 874 \\ \hline 7744 \\ 13522 \\ 15488 \\ \hline 1692064 \end{array}$$

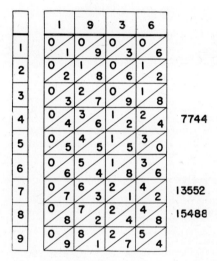

Figure 5.6 Napier's rods (Base-10) set for the multiplicand 1936.

The abacus operation consists in the addition of the three partial products read from the Napier rods. We first set 7744 on the (0,9) abacus, Fig. 5.7a. Then the other two partial products are added, as indicated in (b) and (c). The product 1692064 then appears on the abacus, Fig. 5.7c. The arrangement of the rods in the desired order (Fig. 5.6) takes a little time, of course, but the subsequent procedure is very simple and straightforward.

We have treated four ways of obtaining an abacus product. For particular numbers, there are other shortcuts, as indicated in Chap. 3. In general, however, the combination of abacus and Napier rods is recommended because it requires the minimum number of partial products and eliminates all mental drudgery. Moreover, it is particularly applicable to other number bases.

THE $(0, m)$ ABACUS 113

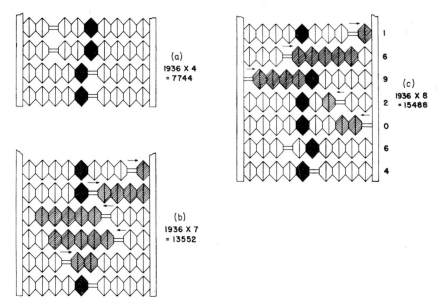

Figure 5.7 The product $1936 \times 874 = 1692064$ obtained on the $(0, 9)$ abacus with Napier's rods.

5.6 Approximation

All of the foregoing products are exact. The product of an s-digit number and a t-digit number will usually have $(s + t)$ digits. But in the great majority of practical cases, such a large number of digits in the product is quite unnecessary and entails considerable waste of effort. If we are working with *experimental data*, there is always a limit to the accuracy of these results. In extreme cases, the data may be accurate to 5 significant figures, but 2 or 3 are much more usual. Suppose that the data are good to 5 digits and that we multiply two such numbers. The product will have 10 digits, but only 5 will have any significance. The other 5 digits are not only meaningless: they are misleading. A similar consideration arises in the computation of mathematical tables, where the number of digits in a product is generally kept the same as the number of digits in each of the factors.

In most cases, a real simplification can be effected by not allowing a product to exceed the desired number of digits. Suppose, for instance, that we want the 4-digit product of

$$8.752 \times 3.614.$$

Moon (0196)

As with any decimal fractions, it is convenient to ignore the decimal point until the final answer is written and then locate it by common sense, as is customarily done with slide-rule calculations. The usual calculation, but moving from left to right on the multiplier, is

$$
\begin{array}{r}
8\ 7\ 5\ 2 \\
3\ 6\ 1\ 4 \\
\hline
2\ 6\ 2\ 5|6 \\
5\ 2\ 5|1\ 2 \\
8|7\ 5\ 2 \\
3|5\ 0\ 0\ 8 \\
\hline
3\ 1\ 6\ 2|9\ 7\ 2\ 8.
\end{array}
$$

We have obtained an 8-digit product, 4 digits of which must be thrown away.

Evidently, the proper thing to do is to compute only as far to the right as necessary. In this example, we take $8752 \times 3 = 26256$; but the other partial products are carried out only as far as necessary. This makes for easy mental

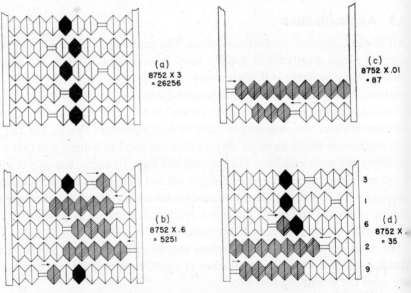

Figure 5.8 The product $8.752 \times 3.614 = 31.63$ correct to 4 digits. Obtained on (0, 9) abacus with Napier's rods.

computation, the whole set-up being

$$\begin{array}{r} 8752 \\ 3614 \\ \hline 26256 \\ 5251 \\ 87 \\ 35 \\ \hline 31629 \end{array}$$

The final product, rounded off to four digits, is

$$8.752 \times 3.614 = 31.63$$

Figure 5.8 indicates the work on the (9,0) abacus.

This method is not limited to the case where the product has the same number of significant digits as the factors. Obvious modifications allow us to cut the product to any desired length. The reader is urged to practice such multiplications on the (9,0) abacus, using the problems of Appendix A.

5.7 Division

Just as abacus multiplication can be closely related to the usual pencil-and-paper multiplication, so abacus division is analogous to the familiar long division. As a simple example of the latter, let us divide 1058 by 23:

$$\begin{array}{r} 46 \\ 23\overline{\smash{\big)}1058} \\ \underline{92} \\ 138 \\ \underline{138} \end{array}$$

On the abacus, the dividend is set on the instrument at or near the top. A rod is then left vacant, and the quotient will appear below this vacant rod. It seems unnecessary to place the divisor on the abacus since the divisor is presumably written on the paper in front of the operator.

In the above example, therefore, 1058 is entered on the abacus, Fig. 5.9a. The operator sees immediately that 23 is divisable into 105 four times. He thus moves 4 beads on the quotient rod (Fig. 5.9b). Since $4 \times 23 = 92$, he subtracts this number from the 105 set on the abacus, which leaves a remainder of 138 in the dividend.

For the second step, the operator ascertains that 23 goes into 138 six times. He sets 6 on the quotient rod below his previous 4. He then subtracts $6 \times 23 = 138$ from the remainder that was on the abacus (Fig. 5.9c), leaving zero. The quotient 46 is then read from the instrument.

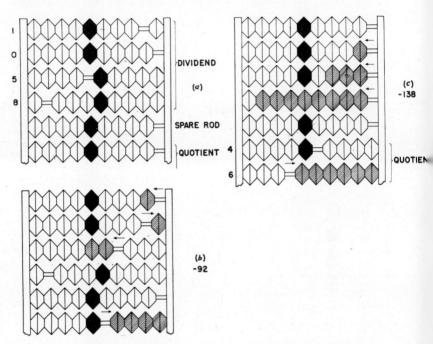

Figure 5.9 The quotient $1058/23 = 46$ on the (0, 9) abacus.

The Napier rods are helpful in division as well as in multiplication. When set up to represent the divisor, they exhibit products of this number by all the digits. For example, let us divide 1692064 by 1936. The customary long division is written

$$\begin{array}{r} 874 \\ 1936 \overline{\smash{)}1692064} \\ \underline{15488} \\ 14326 \\ \underline{13552} \\ 7744 \\ \underline{7744} \end{array}$$

THE (0, m) ABACUS

The operation is very easy on the abacus, aided by Napier's rods. The rods are arranged for the divisor 1936, as shown in Fig. 5.6. The dividend is set on the abacus, Fig. 5.10a. A glance at Fig. 5.6 then shows that the first digit in the quotient is 8 and that $8 \times 1936 = 15488$. Accordingly, an 8 is inserted on the upper rod of the quotient and 15488 is subtracted from the dividend

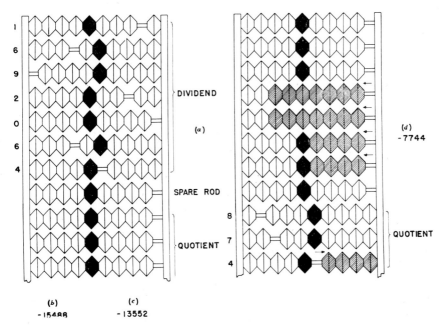

Figure 5.10 The quotient $1692064/1936 = 874$ on the (0, 9) abacus. To save space, the intermediate diagrams (b) and (c) have been omitted.

by moving the proper beads. The abacus now reads 14326. Reference to the arrangement of rods shows that the next digit of the quotient is 7, corresponding to 13552. Subtraction of this number is done as before, leaving 7744 on the abacus. Figure 5.6 shows that this number is exactly divisible by 1936. Thus a 4 is registered in the lower part of the abacus and 7744 is subtracted (Fig. 5.10d).

Of course, most divisions do not have a zero remainder. In such cases, division is continued only until the desired number of digits is obtained. Suppose, for instance, that we wish to evaluate

$$1692164/874$$

to six significant figures. The procedure is similar to that of Figs. 5.10 and 5.06 except that the Napier rods are arranged for 874. The result is 1936.11, which is easily verified on the abacus.

Even with the help of Napier's rods, division on the abacus is much more troublesome than multiplication. The alternative to division consists in multiplication by the reciprocal,[7] A/N being replaced by $A\bar{N}$. If N happens to be a *regular* number N_r, its reciprocal $\bar{N}_r$ is read directly from Table 3.2. If N is an irregular number, its reciprocal is

$$\bar{N} = \bar{N}_r [1 - (\Delta \bar{N}_r) + (\Delta \bar{N}_r)^2 - (\Delta \bar{N}_r)^3 + \cdots], \quad (5.1)$$

where

$$\Delta = N - N_r \quad (5.2)$$

as in Section 3.14. Note that Δ may be either positive or negative. If positive, the terms in the series are alternatively positive and negative; if Δ is negative, all terms in the series are positive. The multiplications and additions of Eq. (5.1) may be done on the abacus. Or, since $\Delta N_r \ll 1$, calculations can often be performed mentally.

For instance, what is the reciprocal of $N = 3996$? From Table 3.2, we select the nearest regular number

$$N_r = 4000, \quad \bar{N}_r = 0.25 \times 10^{-3}.$$

Then $\Delta = 3996 - 4000 = -4$ and

$$\Delta \bar{N}r = -10^{-3},$$

$$(\Delta \bar{N}r)^2 = +10^{-6}.$$

Therefore, the reciprocal of 3996 is

$$\bar{N} = 0.25 \,[1.001001]\, 10^{-3} = 0.25025 \ldots \times 10^{-3}.$$

To evaluate a quotient such as 6821/3996, we merely take the abacus product

$$6821/3996 = 6821 \,(0.25025)\, 10^{-3}$$

$$= 1.70696.$$

5.8 Other Bases

Abaci of type $(0,m)$ are constructed for any base b by using $(b - 1)$ beads per rod. Operation is exactly as with the Base-10 instrument. Abacus *addition* is automatic and does not require the learning of a new addition table for each base.

For instance, suppose that we are operating in the Base-6 system and wish to add *432 + 514*. A (0,5) abacus is employed as indicated in Fig. 5.11. The sum, *1350*, is read from the abacus, Fig. 5.11d.

For *multiplication* and *division*, a new multiplication table is required for each base. If a great amount of computation is to be done with a given base, and the base is not too large, the operator can memorize the new table

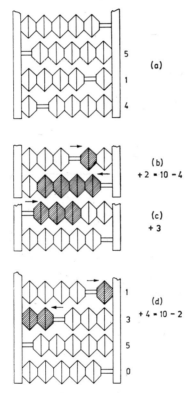

Figure 5.11   The sum *514 + 432 = 1350* on the (0, 5) abacus in the Base-6 system.

without much difficulty. On the other hand, if several bases are being used or if b is large, one has the choice of obtaining the partial products from a table or from Napier's rods (Section 3.11).

Suppose, for example, that the product

$$514 \times 432$$

is to be obtained in the Base-6 system. Rods are assembled for *514* as shown in Fig. 5.12 and partial products are read directly. The product may be written

$$
\begin{array}{r}
514 \\
\times\,432 \\
\hline
1432 \\
2350 \\
3304 \\
\hline
400132
\end{array}
$$

On the (0,5) abacus, we merely add the three partial products that are read from the Napier rods. The first partial product, *1432*, is applied to the abacus (Fig. 5.13a). Then the number *2350* is added (b), remembering in Base-6 that *10* = 6. Finally (c), *3304* is added. The abacus then reads *400132*.

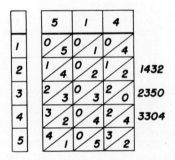

Figure 5.12 Napier's rods for multiplicand *514* in the Base-6 system.

Figure 5.13c of this particular example happens to emphasize a weakness of all minimal abaci. The procedure with a minimal abacus is uniquely dictated by the digits x and y: the operator has no choice. In (c), a *4* must be added on the 3rd rod, but only 2 beads are available. Thus the operator must think $+4 = 10 - 2$. Two beads are moved to the left on the 3rd rod and one bead must be moved to the right on the 4th rod. But no bead is available on the 4th rod, so we must say $+1 = 10 - 5$ and move 5 beads to the left on the 4th rod. This still leaves $3 + 1$ to be added on the 5th rod. And since 4 beads are not available, we must take $3 + 1 = 10 - 2$ and move 2 beads to the left on the 5th rod. Usually such complexity does not occur, though it is always a possibility with a minimal abacus. Nonminimal abaci give much greater flexibility.

THE $(0, m)$ ABACUS

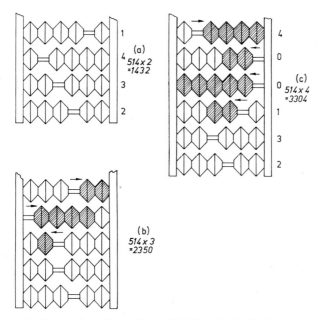

Figure 5.13 The product $514 \times 432 = 400132$ on the $(0, 5)$ abacus in the Base-6 system.

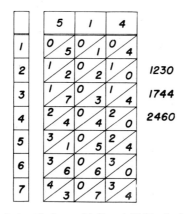

Figure 5.14 Napier's rods for multiplicand 514 in the Base-8 system.

Other bases are handled with equal facility, using the proper abacus (Chap. 4) and Napier rods. Consider, for example, a product of the same numbers in the Base-8 system:

$$\begin{array}{r} 514 \\ \times\,432 \\ \hline 1230 \\ 1744 \\ 2460 \\ \hline 266670 \end{array}$$

Base-8 Napier rods are assembled as shown in Fig. 5.14. From the rods, we read the three partial products which are added on the (0, 7) abacus as shown in Fig. 5.15.

In the Base-12 system, as a final example, we have

$$\begin{array}{r} 514 \\ \times\,432 \\ \hline \varDelta 28 \\ 1340 \\ 1854 \\ \hline 199628 \end{array}$$

The Base-12 Napier rods, assembled to read *514,* give the three partial products *Δ28, 1340,* and *1854.* These three numbers are applied to the (0, 11) abacus to give the product *199628.*

The combination of abacus and Napier rods is also applicable to division in any base. The reader may find it more convenient, however, to divide by using reciprocals (Section 3.14). Equation (5.1) applies in any base, though the operator must remember to do all arithmetic with respect to the correct base. If he is provided with the proper abacus and Napier rods, this requirement presents no difficulties.

As an example, suppose that we wish to divide *531402* by *131* in the Base-6 system and to express the quotient to five significant figures. The closest regular number to *131* is *130,* so

$$N = 131, \quad N_r = 130, \quad \bar{N}_r = 4.0 \times 10^{-3}.$$

THE $(0, m)$ ABACUS

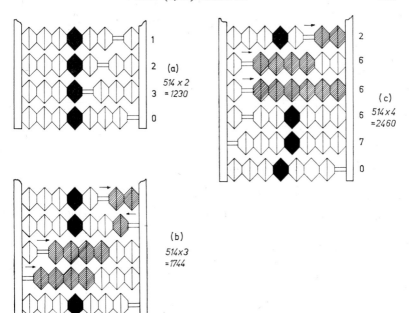

Figure 5.15 The product $514 \times 432 = 266670$ on the (0,7) abacus in the Base-8 system.

Thus

$$\Delta = +1$$

$$\Delta \bar{N}_r = 4 \times 10^{-3}$$

$$(\Delta \bar{N}_r)^2 = 24 \times 10^{-6}.$$

According to Eq. (5.1), the reciprocal of N is

$$\bar{N} = 4\,[1 - 0.004 + 0.000024]\, 10^{-3} = 3.532144 \times 10^{-3}.$$

Therefore,

$$531402/131 = 531402\,(3.532144)\, 10^{-3} = 3344.1.$$

The above operations can be performed very quickly on the $(0, 5)$ abacus. As another example, consider

$$186921/5\varepsilon\varepsilon 9$$

in the Base-12 system. The nearest regular number to *5εε9* is *6000* (see Table 6.1). Therefore,

$$N = 5εε9, \quad N_r = 6000, \quad \bar{N}_r = 0.2 \times 10^{-3},$$

and

$$\Delta = -3,$$

$$\Delta \bar{N}_r = -0.6 \times 10^{-3},$$

$$(\Delta \bar{N}_r)^2 = +0.3 \times 10^{-6}.$$

From Eq. (5.1),

$$\bar{N} = 0.2 \,[1 + 0.0006 + 0.0000003] \, 10^{-3} = 0.20010006 \times 10^{-3}.$$

Thus the quotient is

$$186921/5εε9 = 186921 \,(0.20010006) \, 10^{-3} = 35.18087.$$

The above operations may be done automatically on the $(0,11)$ abacus with Base-12 Napier rods. Or the reader may compare this procedure with ordinary division on the $(0,11)$ abacus.

5.9 Summary

The chapter has examined several ways of handling computations on the $(0,m)$ abacus. The basic operation is, of course, *addition* on the $(0,9)$ instrument (Section 5.1). Of the four methods of *multiplication* considered in this chapter, we recommend the use of an abacus with Napier rods. This combination (Section 5.5) is also helpful in *division*, though here the use of *reciprocals* has advantages (Section 5.7).

The $(0,m)$ abacus is applicable to any base by taking $m = b - 1$. Arithmetic operations are then handled very simply without the memorization of any tables of addition or multiplication (Section 5.8).

In the examples, we have assumed that the abacus is placed on a horizontal desk, with rods running from left to right, and that bead movements are performed with the left hand. If the reader prefers to use his right hand to operate the abacus, he reorients the instrument as in Chap. 2. Such a change causes obvious alterations in the operating instructions, but these alterations should cause the reader no difficulty.

THE $(0, m)$ ABACUS 125

References

1. An exception was the *Millionaire* calculator, made at one time in Switzerland. The multiplication table was built into the machine so that a single turn of the crank would multiply by any digit. Apparently no device of this type is now available. See, for instance;
   F. J. MURRAY, *Mathematical machines*, 2 vols., Columbia Univ. Press, New York, (1961);
   E. M. HORSBURGH, *Napier centenary celebration handbook*, Roy. Soc. of Edinburgh, (1914);
   A. GALLE, *Mathematische Instrumente*, B.G. Teubner, Leipzig, (1912);
   F. A. WILLERS, *Mathematische Maschinen und Instrumente*, Akad. Verlag, Berlin, (1951).
2. Y. YOSHINO, *The Japanese abacus explained*, p. vii, Dover Publications, New York, (1963).
3. T. KOJIMA, *The Japanese abacus*, pp. 88–102, Charles E. Tuttle Co., Rutland, Vt., (1957); Japan Chamber of Commerce and Industry, *Soroban*, Charles E. Tuttle Co., Rutland, Vt., (1967);
   Y. TANI, *The magic calculator*, Japan Publications Trading Co., Tokyo, (1964).
4. A. CUTLER and R. MCSHANE, *The Trachtenberg speed system of basic mathematics*, Doubleday and Co., Garden City, N. Y., (1960).
5. J. NAPIER, *Rabdologia*, Edinburgh, (1617); Leiden, (1626);
   D. E. SMITH, *A source book in mathematics*, p. 182, McGraw-Hill Book Co., New York, (1929);
   J. L. COOLIDGE, *The mathematics of great amateurs*, p. 82, Oxford Univ. Press, (1949);
   K. MENNINGER, *Number words and number symbols*, p. 444, M.I.T. Press, Cambridge, Mass., (1969).
6. A. P. DOMORYAD, *Mathematical games and pastimes*, p. 47, Macmillan Co., New York, (1964).
7. The reader is reminded that the use of reciprocal tables for regular numbers comes from ancient Babylon.
   See O. Neugebauer, *Vorgriechische Mathematik*, Springer-Verlag, Berlin, (1934);
   "Sexagesimalsystem und babylonische Bruchrechnung", *Quellen und Studien*, Springer-Verlag, Berlin, BI, (1931), pp. 183, 452, 458; BII, (1933), p. 199; A3, (1935).

CHAPTER 6

The (*n, m*) Abacus

Of the various forms of abaci, the (*n,m*) is probably the most widely used, including as it does both Chinese and Japanese types.[1] For Base-10, the recommended abacus is (2,5), as indicated in Chap. 4. Most of the present chapter is devoted to this particular instrument (Fig. 6.1).

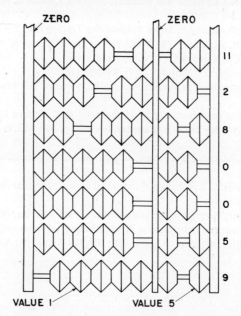

Figure 6.1 The (2, 5) abacus for left-hand operation and with both zeros on the left. The instrument reads 11280059.

The abacus is set on the desk, at the left of the operator and with rods running from left (zero) to right. The zero position for both weights of counters is taken on the left. To set the instrument at zero, therefore, the operator merely tilts it so that all beads slide against the left stops. The abacus is then allowed to fall back to its normal horizontal position on the desk. Manipulation of the beads is done with the index finger and second finger of the left hand, leaving the right hand free for writing.

6.1 Addition

Addition is accomplished solely by movement of the beads, without use of an addition table. By taking similar examples to those employed with the (0,9) abacus, the reader may compare the operation of the two instruments and decide which he prefers. The addition

$$86193 + 12704 = 98897$$

is shown in Fig. 6.2. The number 86193 is first set on the abacus (Fig. 6.2a). We next add 4 on the rod nearest the operator, a 5-bead being pushed to the right by means of the first finger of the left hand (Fig. 6.2b) and a unit bead pushed to the left. The next step is to add 7 on the third rod. This is accomplished by simultaneously sliding a 5-bead with the index finger and two unit beads with the second finger (Fig. 6.2c). The other additions are made similarly. The final result is read from the abacus, Fig. 6.2e.

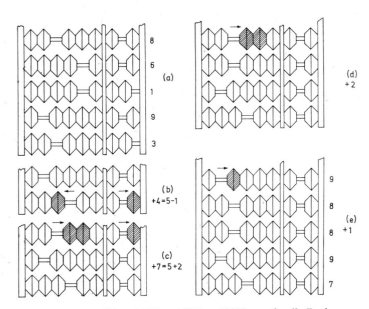

Figure 6.2 The addition $86193 + 12704 = 98897$ on the (2, 5) abacus.

A slightly more complicated addition is indicated in Fig. 6.3, where 47638 is added to 83947. The number 83947 is set on the instrument, Fig. 6.3a. The first step in addition consists in adding 8 on the rod nearest the operator. This may be accomplished by sliding a unit bead on the 2nd rod and subtracting two unit beads on the 1st rod (Fig. 6.3b). Another possibility, of course, is to consider $8 = 5 + 3$ instead of $10 - 2$. Then we move a 5-bead to the right with the forefinger and three unit beads to the right with the second finger. Either operation is permissible, though the second alternative gives a final result that is not quite so easy to read.

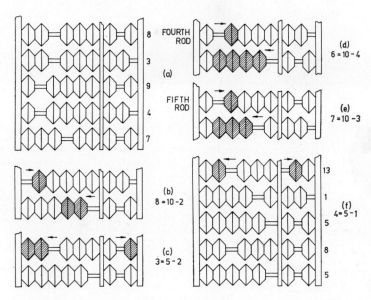

Figure 6.3   The addition $83947 + 47638 = 131585$ on the (2, 5) abacus.

Proceeding with the addition (Fig. 6.3c), we add $3 = 5 - 2$, obtained by moving a 5-bead to the right and two unit beads to the left. Next we add $6 = 10 - 4$, as shown in Fig. 6.3d. An alternate is $6 = 5 + 1$, which can be set on the abacus in an obvious manner. The remaining steps are made similarly, resulting in the final sum of Fig. 6.3f. The abacus is read, of course, by taking the total weight of beads moved to the right for each rod.

6.2 Subtraction

Subtraction on the (2,5) abacus is effected by moving beads to the left. An example is shown in Fig. 6.4, where 47638 is subtracted from 83947. The latter number is set on the abacus, as indicated in Fig. 6.4a. We then perform the operation $-8 = -10 + 2$, Fig. 6.4b, by moving a unit bead to the left

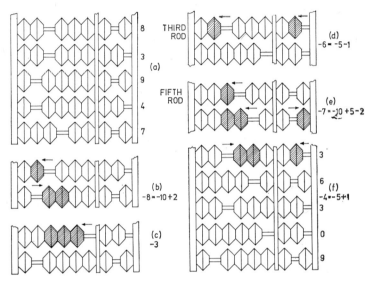

Figure 6.4 The subtraction $83947 - 47638 = 36309$ on the (2,5) abacus.

on the 2nd rod and moving two unit beads to the right on the 1st rod. Thing proceed as usual until (e), which unfortunately requires a triple move: $-7 = -10 + 5 - 2$. Thus we move a unit bead to the left on the fifth rod, a 5-bead to the right on the fourth rod, and two unit beads to the left on the fourth rod. The final result is read from Fig. 6.4f.

6.3 Multiplication

Multiplication on the (2,5) abacus, like multiplication on other Base-10 abaci, requires a knowledge of the ordinary multiplication table. Products of digits must be supplied by the operator, and the function of the abacus is merely to add partial products. If the operator keeps in mind at all times the procedure that he would use in the usual pencil-and-paper method, he should have no difficulty.

9 Moon (0196)

As we saw in Section 3.9, the abacus product of an s-digit multiplicand and a t-digit multiplier involves st partial products. Each possible product of two digits is set on the abacus. The abacus then automatically exhibits the product.

For example, the abacus product 46×23 involves four partial products, which might be written

$$\begin{array}{r} 46 \\ 23 \\ \hline 18 \\ 12 \\ 12 \\ 8 \\ \hline 1058 \end{array}$$

Actually, of course, the partial products are not written but are set immediately on the instrument. The operator thinks $6 \times 3 = 18$ and sets 18 on the abacus. Then he thinks $4 \times 3 = 12$ and sets this number one rod above the

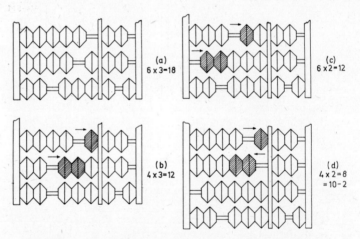

Figure 6.5   The product $46 \times 23 = 1058$ on the (2, 5) abacus.

previous partial product. Finally, $6 \times 2 = 12$ and $4 \times 2 = 8$ are applied to the abacus, which then reads the product 1058. The steps in the process are indicated in Fig. 6.5.

Other products are obtained in a similar manner. The only difference between these products and those of Chaps. 2 and 5 occurs in the addition of

THE (n, m) ABACUS   131

the partial products. The $(2,5)$ abacus tends to give a more flexible addition with less bead movement than the $(0,9)$, usually free from the triple moves associated with the $(1,4)$ instrument.

The disadvantage of the usual method of abacus multiplication outlined above lies in the large number of partial products. By employing Napier's rods in conjunction with an abacus, we reduce the st products to t products. For example, with the product 1936×874, the usual abacus method involves 12 partial products. But with Napier's rods, we have only 3 partial products which are read directly from the rods.

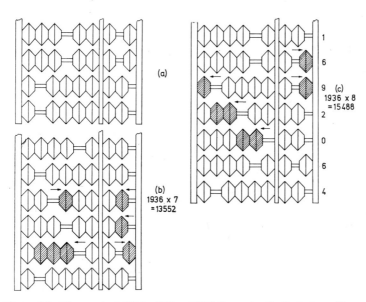

Figure 6.6 The product $1936 \times 874 = 1692064$ on the $(2, 5)$ abacus with Napier's rods.

For the product 1936×874, rods are set up as in Fig. 5.6. The operator then reads from the rods the three partial products

$$1936 \times 4 = 7744,$$

$$1936 \times 7 = 13552,$$

$$1936 \times 8 = 15488,$$

and applies these to the abacus (Fig. 6.6). The instrument then indicates 1692064 as the final product. This procedure corresponds to the familiar scheme:

$$\begin{array}{r} 1936 \\ \times\,874 \\ \hline 7744 \\ 13552 \\ 15488 \\ \hline 1692064 \end{array}$$

It is recommended, therefore, that the (2,5) abacus be used with Napier's rods[2] in performing multiplication and division. The method is a general one that applies to any numbers. The operator should also be on the lookout for special cases that allow shortcuts (Chap. 3).

6.4 Division

The combination of Napier rods and abacus is also helpful in division. The process is closely related to the familiar procedure in long division. Consider, for instance, the quotient 1692064/874. The dividend is first set on the abacus, beginning at the top and leaving space at the bottom for the quotient (Fig. 6.7a). One next sets the Napier rods for 874 (Fig. 6.8). The arrangement

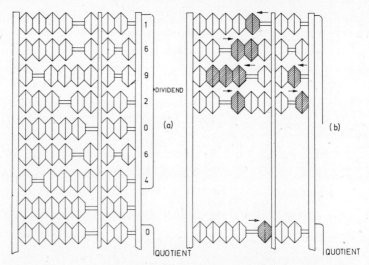

Figure 6.7 The quotient 1692064/874 = 1936 on the (2,5) abacus. The remaining steps are left to the reader.

THE (n, m) ABACUS   133

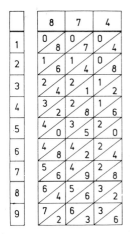

Figure 6.8   Napier's rods for a multiplicand of 874 in the Base-10 system.

of rods allows the operator to see at a glance the successive digits of the quotient. Of course, one can proceed without the rods and *guess* the digits of the quotient, as in ordinary long division. But the rods are less frustrating.

In the above example, the rods show that the first digit of the quotient is 1, corresponding to 874. The next digit, 2, would give 1748, according to Fig. 6.8, which is too large. Thus we set 1 at the lower part of the abacus and subtract 874 in the upper part (Fig. 6.7b). The remainder appears as 8180, which (Fig. 6.8) is over 9 times the divisor. So we set 9 at the bottom of the abacus and subtract 7866. The remainder is now 3146, which gives 3 as the next digit in the quotient. This procedure is continued, giving finally 1936.

Another possibility is to replace division by the multiplication by a reciprocal. The complete list of reciprocals of regular numbers N_r (for $10^5 \leq N_r < 10^6$) is given in Table 3.2. Irregular numbers N are then expressible as

$$N = N_r + \Delta, \qquad (6.1)$$

as discussed in Section 3.14. Here N_r is a value from Table 3.2. The reciprocal of N may be written

$$\bar{N} = \bar{N}_r [1 - (\Delta \bar{N}_r) + (\Delta \bar{N}_r)^2 - (\Delta \bar{N}_r)^3 + \cdots]. \qquad (6.2)$$

Thus a quotient A/N is equal to

$$A/N = A\bar{N}_r [1 - (\Delta \bar{N}_r) + (\Delta \bar{N}_r)^2 - (\Delta \bar{N}_r)^3 + \cdots], \qquad (6.3)$$

which eliminates the troublesome process of division.

134   THE ABACUS

As an example, let us take 1692064/1936. Here $A = 1692064$ and $N = 1936$. From Table 3.2, we select the nearest regular number:

$$N_r = 2000, \quad \bar{N}_r = 0.5 \times 10^{-3}.$$

Then, according to Eq. (6.2),

$$\Delta = -64,$$
$$\Delta \bar{N}_r = -32 \times 10^{-3},$$
$$(\Delta \bar{N}_r)^2 = +1024 \times 10^{-6},$$
$$(\Delta \bar{N}_r)^3 = -32768 \times 10^{-9},$$
$$\bar{N} = 0.5\,[1 + 0.032 + 0.001024 + 0.000033 + \cdots]\,10^{-3}$$
$$= 0.516529 \times 10^{-3}$$

Therefore, the quotient is

$$A/N = 1692.064\,(0.516529) = 874.000$$

which agrees with Fig. 6.6.

6.5 Other Bases

The foregoing treatment of the (n,m) abacus has been restricted to the Base-10 system and to the $(2,5)$ instrument. Obviously, however, the same general methods of abacus calculation can be extended to any base b. The outstanding feature of this change is that each base introduces a new and unfamiliar multiplication table. Of course, the addition table is new also; but this causes no trouble because the abacus takes care of addition automatically. We merely require a new abacus for each base. But for multiplication, a new table must be used for each number system. Since memorization of the multiplication table for a large base is hardly practicable, we find a peculiarly appropriate solution of the difficulty in the combination of Napier's rods and the abacus.[3]

Beyond the decimal system, the most highly advocated base[4] is 12. The use of the dozen and the gross has appealed to Homo sapiens from the earliest times and has been championed by such outstanding men as John Quincy Adams and Herbert Spencer. It was even considered seriously as the base for the metric system of weights and measures. The easy division by 3 has farreaching effects in simplifying arithmetic, and the larger base makes for compact representation of large numbers.

The recommended abacus for Base-12 is the $(2,6)$ shown in Fig. 6.9. The six beads per rod are unit beads, while the two are counters of weight 6. In fact, the $(2,5)$ abacus may be employed with Base-12 by considering the higher beads to have weight 6 instead of 5. Or even the (1,5) may be used as a minimal abacus. But these arrangements do not have the flexibility of the (2,6). The corresponding Napier rods for Base-12 are shown in Fig. 3.5.

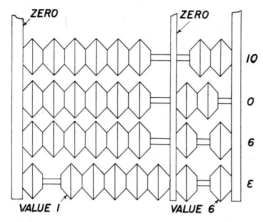

Figure 6.9 The (2, 6) abacus for use with Base-12.

As an example of *addition*, consider

$$\begin{array}{r} \varepsilon 387 \\ 9\varDelta 61 \\ \hline 19228. \end{array}$$

The number $\varepsilon 387$ is first set on the abacus, Fig. 6.10a. We next add *1* on the lowest rod (Fig. 6.10b). The other additions proceed similarly, and the final result is read from the abacus (Fig. 6.10e).

In *multiplication*, we use the (2,6) abacus and the Base-12 Napier rods, Fig. 3.5. For example, consider the product

$$\begin{array}{r} 1936 \\ \times\ 874 \\ \hline 7120 \\ 10506 \\ 12240 \\ \hline 1334180 \end{array}$$

The rods are assembled as shown in Fig. 6.11. By inspection, we see that the pertinent partial products are *7120*, *10506*, and *12240*. These numbers are applied to the abacus as indicated in Fig. 6.12 and the abacus reads *1334180*.

136   THE (n, m) ABACUS

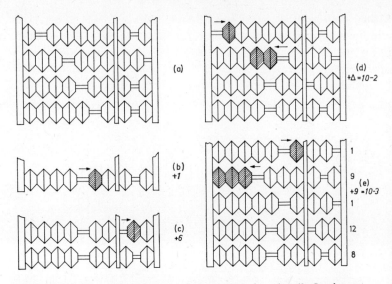

Figure 6.10   The sum $\varepsilon 387 + 9\Delta 61 = 19228$ on the (2, 6) abacus, Base-12.

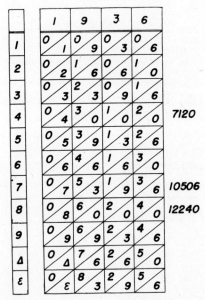

Figure 6.11   Arrangement of Napier rods for multiplicand *1936* in the Base-12 system.

But, one might ask, would it not be simpler to forget about the abacus and the Napier rods and obtain this product by the ordinary pencil-and-paper method shown above? The only trouble with this solution is that the operator must be able to handle mentally both addition and multiplication in the duodecimal system. It is just this difficulty that our procedure attempts to eliminate.

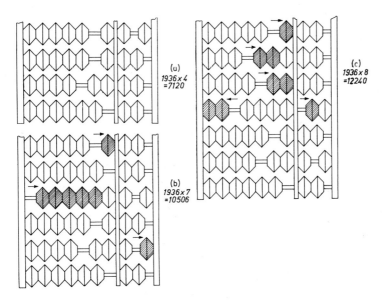

Figure 6.12 The product $1936 \times 874 = 1334180$ on the (2, 6) abacus in the Base-12 system.

Division in Base-12 is the direct analog of Section 6.4 and requires little comment. Perhaps the neatest way is to use reciprocals. The complete set of regular numbers from 10^5 to 10^6 for Base-12 is given in Table 6.1. Equations (6.2) and (6.3) apply, though one must remember to do all calculations in the Base-12 system.

As a simple example, obtain the reciprocal of 3, using Table 6.1. Let

$$N = 3,$$

$$N_r = 4, \quad \bar{N}_r = 0.3.$$

THE ABACUS

Table 6.1 Reciprocals of Regular Numbers
Base-12, $10^5 \leq N_r < 10^6$.

N_r	$\bar{N}_r$	N_r	$\bar{N}_r$
100000	1.0×10^{-5}	368000	0.346×10^{-5}
107854	0.ε483	396900	0.31ε14
116000	0.Δ8	400000	0.3
122800	0.Δ16	426994	0.2Δ209
132300	0.9594	433116	0.2986628
140000	0.9	460000	0.28
151046	0.85176	48Δ800	0.2646
160000	0.8	509000	0.2454
16ε680	0.7716	540000	0.23
183000	0.714	584160	0.2134Δ8
194000	0.69	600000	0.2
1Δ9460	0.63Δ28	63Δ280	0.1Δ946
200000	0.6	690000	0.194
2134Δ8	0.58416	714000	0.183
230000	0.54	771600	0.16ε68
245400	0.509	800000	0.16
264600	0.48Δ8	851768	0.151046
280000	0.46	866230	0.14Δ3314
2Δ2090	0.426994	900000	0.14
300000	0.4	959400	0.1323
		Δ16000	0.1228
31ε140	0.3969	Δ80000	0.116
346000	0.368	ε48300	0.107854

Then $\Delta = -1$ and

$$\Delta \bar{N}_r = -0.3$$
$$(\Delta \bar{N}_r)^2 = +0.09$$
$$(\Delta \bar{N}_r)^3 = -0.023$$
$$(\Delta \bar{N}_r)^4 = +0.0069$$
$$(\Delta \bar{N}_r)^5 = -0.00183$$
$$(\Delta \bar{N}_r)^6 = -0.000509$$
$$(\Delta \bar{N}_r)^7 = -0.0001523$$

Adding these numbers on the abacus in accordance with Eq. (6.2), we obtain

$$\bar{N} = 0.3\,[1 + 0.3 + 0.09 + \cdots] = 0.3\,[1.4000] = 0.40000,$$

which of course is the correct reciprocal in the duodecimal system.

As a second example, consider the quotient $5137/3012$. Here $A = 5137$ and $N = 3012$. The nearest regular number is

$$N_r = 3000, \quad \bar{N}_r = 0.4 \times 10^{-3},$$

according to Table 6.1. Thus

$$\Delta = +12,$$
$$\Delta \bar{N}_r = 4.8 \times 10^{-3},$$
$$(\Delta \bar{N}_r)^2 = 19.94 \times 10^{-6}.$$

By Eq. (6.2),

$$\bar{N} = 0.4 \, [1 - 0.0048 + 0.0000199 - \cdots] \, 10^{-3}$$
$$= 0.3\varepsilon \Delta 5473 \times 10^{-3}.$$

Thus the desired quotient is

$$A/N = 5137 \, (0.3\varepsilon\Delta \, 5473) \, 10^{-3} = 1.8464\varepsilon.$$

Other bases are handled in a similar manner: addition and subtraction on an abacus, multiplication on an abacus with Napier's rods. With Base-16, for instance, one would use a $(2, 8)$ abacus; for Base-30, a $(4, 10)$; for Base-60, a $(10, 10)$.

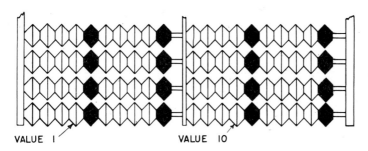

Figure 6.13 A (10,10) abacus for use with Base-60.

The entire chapter has been written on the assumption that the operator is using his left hand for bead-manipulation and that he is taking all zero positions on the left. If he prefers a different convention, however, he may equally well turn the abacus through 90°, use his right hand, and take the zero of the higher beads upward rather than downward. Such changes will cause minor alterations in the operating procedure,—alterations that are easily evaluated by the reader if he should need them.

References

1. K. MENNINGER, *Number words and number symbols*, p. 305, M.I.T. Press, Cambridge, Mass., (1969).
2. D. E. SMITH, *A source book in mathematics*, p. 182, McGraw-Hill Book Co., New York, (1929);
   K. MENNINGER, op. cit., p. 443.
3. A. P. DOMORYAD, *Mathematical games and pastimes*, p. 47, Macmillan Co., New York, (1964).
4. F. E. ANDREWS, *New numbers*, Harcourt, Brace, and World, New York, (1935);
   G. S. TERRY, *Duodecimal arithmetic*, Longmans, Green, and Co., London, (1938).

CHAPTER 7

The (*p*, *n*, *m*) Abacus

We now come to the (p,n,m) abacus mentioned in Section 4.5. The best arrangement of this type for Base-10 is (2,2,4), which will be considered in the first part of the chapter. Later we shall study the $(-4,2,4)$ abacus, where beads of negative value are introduced.

The appearance of the (2,2,4) abacus is shown in Fig. 7.1. On each rod are four unit beads, also two beads of value 2 and two beads of value 5. The purpose of the extra counters is to increase the flexibility of operation. It is a moot question, however, whether the additional complexity resulting from three weights of counters is really advisable and whether the simpler (2,5) is not the better abacus. This is a matter of personal preference.

All zeros may be taken on the left. Thus the abacus is set at zero by temporarily raising the right side, which slides the counters against their left stops. The instrument is then allowed to resume its normal horizontal position on the desk top. Manipulation of the counters is done with the index finger, second finger, and third finger of the left hand. This leaves the right hand free for arranging the Napier rods and for writing numerals.

7.1 Addition

As with other abaci, addition is a mechanical process that requires no knowledge of an addition table. Let us consider the same numerical examples used in the previous chapters. In this way, the reader can, if he is so inclined, make a direct comparison of several types of abacus and decide which he prefers.

An abacus operator on first trying the (2,2,4) will be impressed by the freedom of choice that it allows him. Whereas the (1,4) dictates a unique and inflexible operation for each addition of digits, and the (2,5) gives only a moderate choice, the (2,2,4) allows a freedom that may at first seem almost embarrassing. The aim should be to move a minimum number of counters

WEIGHT 1   WEIGHT 2   WEIGHT 5

Figure 7.1   The (2, 2, 4) abacus for use in the Base-10 system.

and a minimum number of groups. Usually this aim is achieved by the following rules:

a) To add 1 or 3, use unit beads;
b) To add 2 or 4, use beads of weight 2;
c) To add 5, use bead of weight 5.

If these tentative rules are violated, no great harm is done: the result is still correct, though the operation may not have optimum efficiency.

With a minimal abacus, such as the (0,9), the total weight of beads per rod is $(b - 1)$. Thus the digit indicated on any rod can never be as large as the base, and the abacus is read directly without possibility of confusion. With a non-minimal abacus, flexibility of operation is enhanced by having a total weight per rod that exceeds the base. For the (2,2,4), the total weight per rod

THE (p, n, m) ABACUS

is 18 instead of 9. This means that, though the final result is always unambiguous, it may not be as easy to read from the instrument as with a minimal abacus. The following numerical examples will clarify this characteritsic of the $(2,2,4)$ abacus.

Let us add 83947 and 47638. The larger of the two numbers is set on the abacus, as shown in Fig. 7.2a. To add 8 on the lowest rod, the operator thinks $8 = 10 - 2$ and moves a unit counter on the second rod and a 2-bead

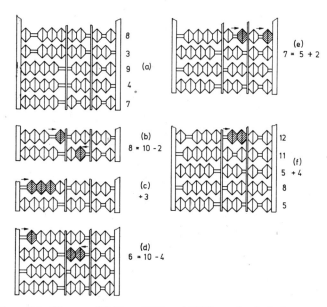

Figure 7.2 The sum $83947 + 47638 = 131585$ on the $(2, 2, 4)$ abacus.

on the first rod (Fig. 7.2b). He next adds 3 by moving three unit beads (Fig. 7.2c). Further additions lead to the final sum (Fig. 7.2f) which reads

$$12, 11, 5, 8, 5$$

which is interpreted as

$$120000 + 11000 + 500 + 80 + 5 = 131585.$$

7.2 Multiplication

Multiplication can be performed directly on the $(2,2,4)$ abacus by adding st partial products. Or multiplication can be effected by the combination of abacus and Napier's rods, using t partial products. Except for differences in the abacus addition, caused by peculiarities in the bead arrangement of the (2,2,4), the process is exactly as in Chaps. 5 and 6.

For example, with the product 1936 × 874, the partial products 7744, 13552, and 15488 are read directly from Napier's rods. The first of these is applied to the abacus, as shown in Fig. 7.3a. The next partial product is added, beginning with the second rod, Fig. 7.3b. Finally, the third partial product, 15488, is added (Fig. 7.3c). Note the simplicity of the bead movements and the small number of beads that are moved.

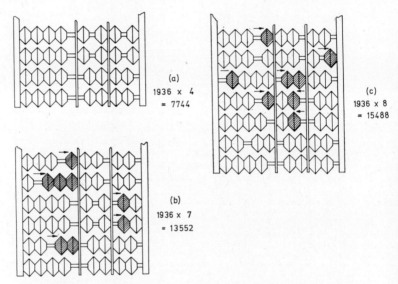

Figure 7.3   The product 1936 × 874 = 1692064 on the (2, 2, 4) abacus.

On the other hand, two of the rods have final sums above 9; but there is no difficulty in seeing that the final result is 1692064. This accumulation of 5-beads to the right on the third and fourth rods could have been eliminated, of course, by a slight change in procedure. In (b), instead of adding 5 by direct movement of 5-beads on third and fourth rods, we could have said +5 = 10 − 5, resulting in the movement of 5-beads to the *left*. The text move, however, is preferable.

Division can be handled by use of a reciprocal table, as explained in Section 3.14. The (p,n,m) abacus can also be applied to other *number bases* (Section 4.8). Further comment seems hardly necessary.

7.3 The (−m, n, m) Abacus

The best $(-m,n,m)$ abacus for Base-10 is probably the $(-4,2,4)$ shown in Fig. 7.4. The two sections on the left constitute a regular $(2,4)$ abacus with two counters of weight 5 per rod and four unit counters. To this arrangement is added the section on the right, containing four negative unit beads per rod.

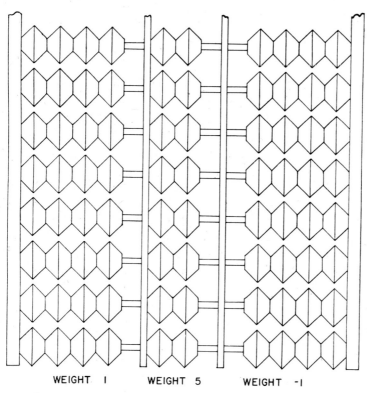

WEIGHT 1   WEIGHT 5   WEIGHT −1

Figure 7.4 The $(-4, 2, 4)$ abacus.

The zeros for the first two of these sections are to the left, but the zero for the negative section is on its right. This scheme means than an *increase* is always effected by moving counters to the right, while a *decrease* is always obtained by bead movement to the left. To set the instrument at zero, the operator raises the right edge, allowing all beads to slide to the left. After the abacus is again resting horizontally on the table, a finger can be drawn across the rods to move the negative beads against their right stop.

Moon (0196)

Operation of the $(-4,2,4)$ abacus is very similar to operation of the (n,m). Occasionally, however, one finds sums of digits that are simplified if the negative beads are employed. As indicated in Chap. 4, the purpose of the negative counters is to allow full use of complements. Thus our rule for addition of x and y on the $(-4,2,4)$ abacus is to *add directly* if $y \leq 5$ but to *use the complement* if $y > 5$. This provides an invariable procedure, irrespective of x, and it tends to reduce bead movement to a minimum.

For example, let us add 83947 and 47638 on the $(-4,2,4)$ abacus. The number 83947 is set on the abacus, Fig. 7.5a. As a first step in addition, we introduce 8; and since this number is above 5, we add 10 and subtract the

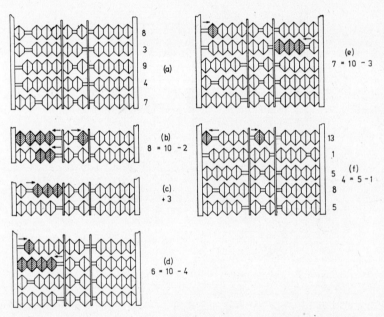

Figure 7.5 The sum $83947 + 47638 = 131585$ on the $(-4,2,4)$ abacus.

complement. In this particular case, the complement is most easily subtracted by moving two positive unit counters to the left (Fig. 7.5b). The negative counters are available, however, and the operator may move two of them to the left if he wishes, instead of moving the positive counters.

The number 3 is introduced in the usual way (Fig. 7.5c), but 6 is added by employing its complement. Here four positive unit beads are moved to the left (Fig. 7.5d). Another possibility is to move four negative beads to the left

THE (p, n, m) ABACUS 147

Figure 7.5e shows a case where negative beads are moved. Here 7 is to be added, which is effected by moving a unit bead to the right on the fifth rod and three negative beads to the left on the fourth rod. Positive unit beads could have been moved equally well; but if three positive beads had not been available, the negative beads would have been really useful. In (f), we have the final sum 131585.

As noted above, the particular example of Fig. 7.5 does not really need the negative beads. An example that better utilizes the potentialities of the negative counters is the addition of 5126 and 4768. The number 5126 is set on the abacus, Fig. 7.6a. The first step (Fig. 7.6b) consists in adding 8; and since it is greater than 5, our rule requires the use of the complement, or $+8 = 10 - 2$. But there is only one positive unit counter available on the first rod, so we are forced to move two negative counters.

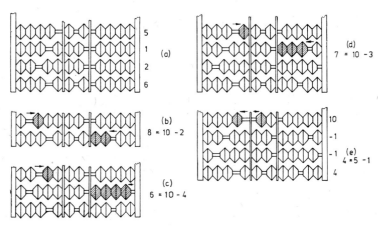

Figure 7.6 The sum 5126 + 4768 = 9894 on the $(-4,2,4)$ abacus.

Similarly in Fig. 7.6c we add $+6 = 10 - 4$. Again there are not enough positive counters so we move four negative counters to the left. Also in Fig. 7.6d, we are forced to move three negative counters. The final result appears in (e) as

$$10, -1, -1, 4$$

obtained by mentally adding for each rod the weights of all positive counters not at zero and subtracting the number of negative counters not at zero. But the above is equivalent to

$$10000 - 100 - 10 + 4 = 9894.$$

10a Moon (0196)

148   THE ABACUS

In this example, the final reading of the abacus is not immediately evident in its desired form. This is an extreme illustration of the non-unique nature of counter arrangement in a non-minimal abacus. Fortunately, the readings of the $(-4,2,4)$ abacus are usually more easily interpreted than in this illustration.

7.4 Products on the $(-m, n, m)$

Multiplication on the $(-4,2,4)$ abacus is most easily accomplished by the help of Napier's rods. The partial products are obtained exactly as with the previous abaci, and the only difference introduced by the $(-4,2,4)$ is a slight change in the addition of these partial products.

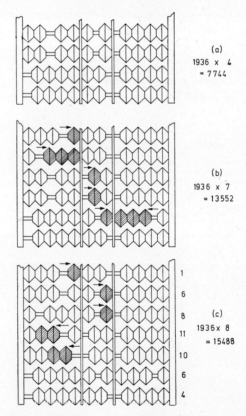

Figure 7.7   The product $1936 \times 874 = 1692064$ on the $(-4, 2, 4)$ abacus.

THE (p, n, m) ABACUS

Taking our usual example, 1936×874, we use Napier's rods to obtain the products 7744, 13552, and 15488. These numbers are added on the $(-4, 2, 4)$ instrument as indicated in Fig. 7.7. The first partial product is set on the abacus (Fig. 7.7a). The second and third are then added (Figs. 7.7b and c). The final product appears as 1, 6, 8, 11, 10, 6, 4 which is easily interpreted as 1692064.

Division on the $(-m, n, m)$ abacus introduces nothing new and will not be elaborated here. As noted in Chap. 4, negative beads are applicable to other *number bases*. Details of such applications are obvious. The foregoing chapter has assumed that the operator prefers to manipulate the beads with his left hand and that he places the abacus on his left with rods running from left (zero) to right. Only minor changes in the instructions, however, are needed if the abacus is reoriented and operated with the right hand or if the zeros are relocated. Such changes can be made at the option of the reader and need cause no change in the design or construction of the abacus.

We have now completed our treatment of the abacus, having sketched its history and discussed abacus design and operation. Evidently the subject is a much bigger one than would have been expected for such a lowly tool. I only hope that the reader gets as much fun out of it as I have!

APPENDIX A

Exercises in Abacus Arithmetic*

Addition

(1) 431 + 343 =
(2) 998 + 453 =
(3) 968 + 639 =
(4) 6653 + 935 =
(5) 7055 + 1216 =

(6) 9026 + 8139 =
(7) 5302 + 4129 =
(8) 6985 + 1427 =
(9) 3167 + 2746 =
(10) 7727 + 7134 =

(A) Addition of 2-digit numbers

(11)	(12)	(13)	(14)	(15)
67	59	16	50	38
88	51	12	89	18
59	39	22	17	12
92	69	75	50	86
95	60	35	44	73
47	31	52	84	93
27	71	38	25	82
13	80	97	77	26
86	89	69	33	57
55	74	70	15	83

(B) Addition of 3-digit numbers

(16)	(17)	(18)	(19)	(20)
817	880	245	914	314
462	885	240	966	924
999	293	528	473	773
181	834	330	507	431
969	238	460	198	343

---

* Based on Rand Corp.. *A million random digits*, Free Press, Glencoe, I 11.. (1955).

APPENDIX A

(16)	(17)	(18)	(19)	(20)
427	620	323	831	311
902	639	998	453	800
968	935	423	975	203
625	232	569	300	219
610	534	294	568	707

(C) Addition of 4-digit numbers

(21)	(22)	(23)	(24)	(25)
1561	4979	9182	4583	8510
8179	4504	1181	2083	1767
4152	9287	7042	1116	7183
5944	4398	1258	5010	7159
3277	6354	9239	4475	4737
1653	4054	2884	1340	6997
4005	6653	7546	4805	3741
2911	3311	2338	2746	8701
8742	5077	7834	7031	9585
2472	7727	7803	7439	9691

(26)	(27)	(28)	(29)	(30)
6098	1177	7447	3304	5788
4763	4867	3875	9026	8139
4213	4689	1317	2241	3838
6488	6985	6906	2068	5548
5250	6020	1427	8705	6836
2832	3781	7291	4129	2019
3483	8809	4325	3167	5302
7945	7211	1701	5284	7055
1413	3619	8049	2258	4845
6277	5583	7910	4414	3364

(D) Addition of 5-digit numbers

(31)	(32)	(33)	(34)	(35)
56316	51632	12874	77884	64611
37723	54799	82160	70034	19736
21424	27973	67202	16071	25589
26664	68568	85199	53362	46569
63804	68465	27908	28119	45206

(31)	(32)	(33)	(34)	(35)
75139	98500	67022	56786	48215
36534	28681	49810	30883	69523
18579	18369	77929	28540	17423
90833	24279	96212	76029	91807
98849	96335	81153	37074	90039

(36)	(37)	(38)	(39)	(40)
30393	61477	84746	38688	56405
58319	89731	28302	39968	17839
85098	18421	13264	32604	92073
66519	29861	70595	11694	57622
57571	52829	12933	46262	93328
24541	80038	46831	73262	15442
35062	78040	24864	45405	50186
14400	43350	47275	43923	75070
62731	74323	20527	67397	58001
82534	82892	70196	88228	30383

Subtraction

(41) 56316 − 37723 =
(42) 75139 − 63804 =
(43) 98500 − 28681 =
(44) 67022 − 49810 =
(45) 53362 − 28119 =
(46) 90039 − 17423 =
(47) 85098 − 66519 =
(48) 78040 − 52829 =
(49) 28302 − 13264 =
(50) 73262 − 50186 =

Combined addition and subtraction

(51)	(52)	(53)	(54)	(55)
56316	51632	12874	77884	64611
37723	54799	82160	−70034	−19736
−21424	−27973	67202	16071	−25589
26664	68568	−85199	53362	46569
63804	68465	27908	−28119	45206

(51)	(52)	(53)	(54)	(55)
−75139	−98500	−67022	56786	−48215
−36534	−28681	49810	−30883	69523
18579	18369	−77929	28540	−17423
90833	−24279	96212	−76029	91807
−98849	96335	−81153	37074	90039

Multiplication

It is suggested that the reader try these exercises also with Napier rods.

(A) Multiplication of 2-digit numbers. Obtain exact product

(56) $92 \times 59 =$
(57) $86 \times 47 =$
(58) $89 \times 71 =$
(59) $75 \times 52 =$
(60) $69 \times 12 =$
(61) $84 \times 33 =$
(62) $83 \times 26 =$
(63) $89 \times 18 =$
(64) $97 \times 22 =$
(65) $69 \times 47 =$

(B) Multiplication of 3-digit numbers

Obtain exact product.

(66) $817 \times 462 =$
(67) $831 \times 453 =$
(68) $924 \times 507 =$
(69) $534 \times 232 =$
(70) $601 \times 473 =$
(71) $707 \times 219 =$
(72) $885 \times 245 =$
(73) $967 \times 462 =$
(74) $935 \times 620 =$
(75) $814 \times 187 =$

(C) Multiplication of 4-digit numbers

Obtain product correct to 4 significant figures.

(76) $0.8179 \times 1561 =$
(77) $0.4398 \times 1653 =$
(78) $2911 \times 0.1653 =$
(79) $7727 \times 0.7439 =$
(80) $8701 \times 0.3741 =$
(81) $0.4583 \times 1116 =$
(82) $7042 \times 0.1258 =$
(83) $4054 \times 6.354 =$
(84) $91.82 \times 1767 =$
(85) $7.159 \times 50.10 =$

(D) Multiplication of 5-digit numbers

Obtain product correct to 5 significant figures.

(86) $0.61478 \times 30393 =$
(87) $35062 \times 0.24541 =$
(88) $4.3350 \times 14.400 =$
(89) $0.74323 \times 205.27 =$
(90) $75.070 \times 0.15442 =$
(91) $0.46569 \times 45.206 =$
(92) $77884 \times 0.53362 =$
(93) $85199 \times 6.8579 =$
(94) $2666.4 \times 2.1526 =$
(95) $18.579 \times 0.17839 =$

Division

(A) Obtain exact quotients

(96) $612296/8 =$
(97) $637962/78 =$
(98) $4485680496/76537 =$
(99) $5428/59 =$
(100) $4042/86 =$
(101) $6319/71 =$
(102) $3900/52 =$
(103) $377454/462 =$
(104) $284273/601 =$
(105) $152218/814 =$

APPENDIX A 155

(B) Obtain 3 significant figures in quotient

(106) 35857/58 =   (116) 47280.1/408 =
(107) 81569/87 =   (117) 23.245/5.38 =
(108) 67626/0.915 =   (118) 58283/153 =
(109) 52652/272 =   (119) 19407/162 =
(110) 59563/5.36 =   (120) 37915/9.20 =
(111) 79834/27 =   (121) 137.595/0.309 =
(112) 44965.7/55 =   (122) 71852/415 =
(113) 72241/0.904 =   (123) 20811/5.77 =
(114) 15980/71.5 =   (124) 98649/88.5 =
(115) 39.625/0.971 =   (125) 56.710/1.053 =

(C) Obtain 5 significant figures in the quotient

(126) 52478/22835 =   (136) 941.32/1.5190 =
(127) 33307/73842 =   (137) 840.25/702.98 =
(128) 67277/32880 =   (138) 22.002/8.0519 =
(129) 76457/94489 =   (139) 2351.6/0.86294 =
(130) 82597/40836 =   (140) 32871/89573 =
(131) 80249/16089 =   (141) 56605/86696 =
(132) 10964/21414 =   (142) 37.707/9.0117 =
(133) 72117/91712 =   (143) 1.7511/2.7701 =
(134) 11487/67479 =   (144) 357.64/0.88217 =
(135) 13649/94539 =   (145) 7050.5/7.530 =

Addition

6th-Grade Operator

The following tests are similar to the examinations given in Japan in licencing abacus operators. The 10 columns are to be summed (70 percent accuracy) in 10 minutes.

(146)	(147)	(148)	(149)	(150)
492	338	233	575	410
692	910	−187	175	117
721	386	989	525	593
246	709	570	650	722
905	925	550	281	−767
371	064	817	515	501

156   THE ABACUS

(146)	(147)	(148)	(149)	(150)
606	768	−172	839	209
425	201	883	687	593
427	845	330	905	−849
460	904	−271	225	753
495	590	158	422	638
725	146	211	720	829
170	711	−499	533	−961
207	344	988	920	669
679	192	174	483	803

(151)	(152)	(153)	(154)	(155)
738	938	591	409	202
898	773	419	984	792
203	−304	558	−413	741
699	564	589	716	410
531	882	552	144	589
230	666	972	−663	398
537	−601	475	440	602
123	720	932	915	300
748	670	527	100	317
861	364	848	−881	679
590	−456	800	364	561
467	760	581	182	856
628	803	595	116	738
360	996	601	−294	503
538	228	481	765	321

10 columns to be summed (70% accuracy) in 10 minutes.

(156)	(157)	(158)	(159)	(160)
3100	870	472	3389	26
638	14	93	889	87
371	83	97	236	795
78	331	−203	60	856
130	5636	3580	6100	−65
8076	432	469	688	22
191	76	−88	48	72
64	8856	64	806	5165
95	800	749	74	−107

APPENDIX A

(156)	(157)	(158)	(159)	(160)
508	88	49	87	52
17	44	4574	78	9946
51	598	−458	642	31
329	44	42	425	−521
41	68	672	92	628
66	794	21	91	719

(161)	(162)	(163)	(164)	(165)
539	97	856	1399	896
848	80	83	574	309
7855	903	79	60	45
707	−548	483	−908	75
53	57	7485	43	645
8038	421	88	857	1780
67	−143	74	61	897
860	1160	915	−890	408
309	530	77	37	61
24	5873	856	13	50
47	95	1270	567	5641
696	−780	51	6708	528
78	96	111	−947	79
32	62	504	43	42
63	87	90	57	796

5th-Grade Operator

10 columns to be summed (70% accuracy) in 10 minutes.

(166)	(167)	(168)	(169)	(170)
2035	1349	684	6148	718
508	674	194	109	775
661	367	9317	−786	3924
110	3270	303	807	906
9219	−187	406	689	306
682	819	274	3715	2975
842	290	4452	4384	644
872	509	147	−768	704
3597	−905	982	901	6832
482	9997	634	5124	405

(166)	(167)	(168)	(169)	(170)
467	−655	403	935	912
4325	9374	7321	−382	3477
663	4520	8727	6777	210
2090	487	882	606	633
155	750	6143	271	1229

(171)	(172)	(173)	(174)	(175)
992	837	5612	138	856
888	−248	646	105	302
6530	968	618	1503	279
868	2539	877	551	4816
193	−504	809	2222	231
8955	1500	199	−997	902
9174	1602	4383	825	7522
489	383	842	5622	5227
632	862	704	697	384
646	−882	791	−549	310
2384	7468	915	8564	221
986	301	5018	852	7454
8599	−996	903	5877	3025
407	7340	4553	566	203
639	945	5458	−737	232

10 columns to be summed (70% accuracy) in 10 minutes.

(176)	(177)	(178)	(179)	(180)
933	7035	2801	8074	393
5028	857	905	8663	8072
643	−1824	831	1179	397
3909	574	9402	−421	8810
7009	7308	8570	202	6078
1469	−3070	4175	6850	516
844	542	137	−1404	775
6045	2593	551	2207	6580
3208	2162	2899	368	4244
188	982	785	−2679	2185
8752	352	3346	524	2775
287	1724	7338	358	4908

APPENDIX A 159

(176)	(177)	(178)	(179)	(180)
2294	−5411	3404	7896	2544
6560	6409	8600	9397	5500
5002	8169	5576	4623	865

(181)	(182)	(183)	(184)	(185)
342	510	2487	5630	510
237	1008	6931	790	8300
2744	−201	987	9011	619
8392	539	7444	−8117	5054
6400	1437	5378	3650	825
7880	4994	248	501	4311
791	202	8666	5986	5175
427	8926	5614	8997	8242
2368	600	114	−963	8006
3404	−7225	1090	3840	374
7452	1801	637	823	1814
174	4710	6367	−962	688
4369	6350	543	4755	7131
2899	−1114	5544	5734	6884
1312	8890	9408	8320	3249

10 columns to be summed (70% accuracy) in 10 minutes.

(186)	(187)	(188)	(189)	(190)
2487	4940	5630	483	510
693	842	790	200	8300
198	−923	901	9322	619
7744	2816	181	−1345	505
453	209	173	784	482
782	−789	6505	931	5431
4886	1515	150	5470	1517
665	566	986	−1148	582
6141	742	8997	780	428
1410	667	963	542	6003
906	8579	3840	−265	741
376	−605	823	3588	814
367	935	9624	563	6887
543	7319	755	631	131
554	853	573	948	688

(191)	(192)	(193)	(194)	(195)
4324	557	3955	203	927
971	−374	1380	6911	886
593	546	5810	−289	2546
8559	2124	445	994	2614
559	8464	1846	993	783
834	858	380	2922	529
140	4358	812	884	1001
8806	−1062	5124	−2990	277
917	2007	322	484	611
328	507	351	420	789
5880	391	632	1053	258
134	−193	669	−308	6071
108	802	405	3320	663
3225	499	706	636	443
905	720	573	137	5057

4th-Grade Operator

10 columns to be summed (70% accuracy) in 10 minutes.

(196)	(197)	(198)	(199)	(200)
1265	3673	2132	9075	7968
1616	7988	5157	9644	6179
4611	−6377	3224	−1054	6976
7697	2407	1273	1796	6018
5109	6251	7431	6075	3748
8699	6005	9702	6105	5592
6976	4646	3635	1753	8036
6925	8809	2973	−8551	1241
7573	3437	9779	3780	5408
2535	2078	5218	8360	6492
7012	−6733	9607	9594	6404
2767	1987	4421	6955	7605
6381	2982	3874	4755	6870
7697	8615	1464	−6839	3330
4436	−4371	7887	8078	2001

(201)	(202)	(203)	(204)	(205)
1404	6236	1546	7462	4110
2535	8626	9905	6683	1173

APPENDIX A

(201)	(202)	(203)	(204)	(205)
3607	−2362	7478	9488	3648
7790	7421	3036	5627	9515
4157	4891	6885	−7405	3674
4117	3924	1393	2370	1788
2364	−7522	6421	1456	4052
7342	4298	1771	−3065	6008
1234	8326	8312	8915	5754
3929	9533	2961	8220	9321
5053	9117	9422	3544	3619
3334	4579	5318	2604	1170
3759	−7957	1618	−1411	9532
9110	5299	6252	1191	1239
8256	7751	3674	6519	1576

10 columns to be summed (70% accuracy) in 10 minutes.

(206)	(207)	(208)	(209)	(210)
43398	5448	67297	78207	66199
7966	9525	2516	34228	9981
6376	−3101	3163	3944	1138
87364	5000	8662	−6077	8731
3134	67002	6870	6580	3014
82200	31464	7398	6270	1034
7793	4973	18963	8869	63717
8828	3925	6003	5501	10310
71137	76581	5355	42421	3360
85054	5139	9814	−56523	70185
6466	−29770	40414	7821	8282
7087	8117	8725	3999	6691
1115	32440	6959	5084	98773
2704	−7081	16184	−7528	4912
2232	4214	32709	10328	2639

(211)	(212)	(213)	(214)	(215)
8012	41686	24148	5414	3637
5849	8548	9524	7016	9135
8447	7915	7123	3895	88445
80490	−2628	4039	−9546	87310

(211)	(212)	(213)	(214)	(215)
9081	5915	4077	6666	1534
1106	46357	1596	30544	9712
43995	−7149	53961	67089	5146
83182	59117	61089	40524	33188
3964	3029	6869	−1925	5201
7415	16826	9388	1574	8171
4203	66576	9187	4069	57656
2866	−8096	1532	1000	6338
9743	7755	29896	−7529	6537
16618	3781	4262	10668	3761
59364	9990	12716	23743	61363

10 columns to be summed (70% accuracy) in 10 minutes.

(216)	(217)	(218)	(219)	(220)
292	7262	23931	33595	10716
5736	22555	55916	8370	3552
600	−1382	484	−792	173
810	844	77330	545	850
8067	8555	6776	8386	99275
4932	316	5728	6068	97475
3666	−852	483	−952	11064
22251	57931	5962	8019	934
17616	840	806	8054	920
6071	6357	8558	70427	5362
677	5164	23560	57927	57562
125	−652	8924	−728	995
18653	94703	59644	833	8281
40426	60375	313	87400	4658
95304	50816	697	72301	9005

(221)	(222)	(223)	(224)	(225)
7770	18418	4373	20660	654
6535	1682	489	70109	3321
751	−6263	733	−10273	7353
6463	5551	87697	4263	64444
30050	8410	28098	377	5069
44697	12351	70926	−445	32023

APPENDIX A 163

(221)	(222)	(223)	(224)	(225)
94868	59866	227	2403	60078
2269	−5898	907	903	3104
7337	741	9293	144	59967
406	815	10093	12100	9762
700	13911	7267	1297	538
10403	13130	36025	−1595	1056
475	220	789	74482	906
622	−551	9818	101	383
35864	273	8072	877	932

3rd Grade Operator

10 columns to be summed (70% accuracy) in 5 minutes.

(226)	(227)	(228)	(229)	(230)
303462	9678	25893	72872	37009
6957	9225	57092	541	521733
7824	5112	763967	907	6327
403	−11186	2486	462622	72135
195	683	62423	−111872	33005
248	622705	27618	86467	28701
3732	80366	8418	72422	34710
51170	−419	4789	436804	4935
88022	290	227	−644	950
700526	611151	35615	2504	693
226188	450174	281813	2439	893112
11405	−81263	902	2072	596170
657841	315075	508	−41994	386
83300	47189	174640	434450	7418
88000	99951	917469	81974	577536

(231)	(232)	(233)	(234)	(235)
7396	15902	31694	1729	890
247	613639	5563	2184	1386
2201	5268	3188	3005	492
2296	41377	1390	959338	75047
416	25684	291	365375	59643
23628	−801	50275	−8165	310743
73379	516	12230	624	8172

(231)	(232)	(233)	(234)	(235)
432	1816	79607	−1081	307
558409	58555	954	38012	1832
60963	54305	263	412302	119695
36047	−861	49000	502875	66071
270631	8953	977823	−46599	43445
647637	845178	106965	83785	177837
29082	519260	673884	141	307529
260280	−309091	269003	211	52606

(231)	(232)	(233)	(234)	(235)
7396	15902	31694	1729	890
247	613639	5563	2184	1386
2201	5268	3188	3005	492
2296	41377	1390	959338	75047
416	25684	291	365375	59643
23628	−801	50275	−8165	310743
73379	516	12230	624	8172
432	1816	79697	−1081	307
558409	58555	954	38012	1832
60963	54305	263	412302	119695
36047	−861	49000	502875	66071
270631	8953	977823	−46599	43445
647637	845178	106965	83785	177837
29082	519260	673884	141	307529
260280	−309091	269003	211	52606

10 columns to be summed (70% accuracy) in 5 minutes.

(236)	(237)	(238)	(239)	(240)
83183	44799	61330	84445	11364
15461	33555	11838	77205	46345
88997	58608	37496	84394	40639
96634	55270	74484	40760	19572
39343	41848	83272	73845	34159
76187	27512	89275	17361	12518
51649	89046	10818	67790	86926
69036	61975	38111	10353	65650
40387	79250	87939	36885	14931
43933	65479	44211	34317	57011

APPENDIX A 165

(236)	(237)	(238)	(239)	(240)
86561	21693	45991	44264	13487
67600	78499	21942	62994	32387
87081	77459	34406	23179	76475
76544	73214	28785	86523	72583
89082	50062	41740	35982	57269

(241)	(242)	(243)	(244)	(245)
20420	23375	81683	47740	13830
57224	29913	54726	49996	51094
70061	24245	46546	90997	31691
28379	78402	95474	40690	97311
21115	37901	54716	73062	47805
17471	21882	40624	99417	85552
44765	77019	97378	84362	39430
26548	79658	15645	36977	82075
66533	47396	87183	56369	29116
61231	86300	88180	33825	76537
65829	79409	44776	76063	95406
31960	54902	41489	24841	20908
22771	27283	12313	77021	86244
61051	19483	88860	90894	47511
34509	87369	90769	16615	92035

Multiplication

6th-Grade Operator

5 minutes for each group of 10 (70% accuracy).

(246) $96 \times 874 =$
(247) $98 \times 631 =$
(248) $54 \times 645 =$
(249) $81 \times 170 =$
(250) $98 \times 613 =$
(251) $18 \times 448 =$
(252) $39 \times 298 =$
(253) $48 \times 702 =$
(254) $50 \times 232 =$
(255) $22 \times 703 =$

(256) $56 \times 4451 =$
(257) $30 \times 4493 =$
(258) $37 \times 9297 =$
(259) $23 \times 1436 =$
(260) $16 \times 8084 =$
(261) $99 \times 2062 =$
(262) $79 \times 5277 =$
(263) $17 \times 5386 =$
(264) $94 \times 3195 =$
(265) $57 \times 8975 =$

166   THE ABACUS

5th-Grade Operator

5 minutes for each group of 10 (70% accuracy).

Group A		Group B	
(266)	$510 \times 468 =$	(276)	$990 \times 292 =$
(267)	$182 \times 913 =$	(277)	$605 \times 269 =$
(268)	$385 \times 182 =$	(278)	$622 \times 858 =$
(269)	$401 \times 468 =$	(279)	$842 \times 259 =$
(270)	$267 \times 941 =$	(280)	$364 \times 374 =$
(271)	$674 \times 637 =$	(281)	$367 \times 485 =$
(272)	$180 \times 781 =$	(282)	$264 \times 286 =$
(273)	$211 \times 826 =$	(283)	$285 \times 969 =$
(274)	$790 \times 629 =$	(284)	$307 \times 671 =$
(275)	$562 \times 734 =$	(285)	$760 \times 996 =$

4th-Grade Operator

5 minutes for each group of 10 (70% accuracy).

Group A		Group B	
(286)	$61 \times 77725 =$	(296)	$88 \times 21237 =$
(287)	$41 \times 78033 =$	(297)	$52 \times 19013 =$
(288)	$81 \times 67276 =$	(298)	$65 \times 92901 =$
(289)	$77 \times 66964 =$	(299)	$96 \times 47223 =$
(290)	$37 \times 84946 =$	(300)	$17 \times 98508 =$
(291)	$87 \times 31998 =$	(301)	$22 \times 37282 =$
(292)	$76 \times 30684 =$	(302)	$99 \times 90043 =$
(293)	$82 \times 11592 =$	(303)	$82 \times 54528 =$
(294)	$31 \times 51998 =$	(304)	$83 \times 57667 =$
(295)	$53 \times 92001 =$	(305)	$21 \times 67588 =$

3rd-Grade Operator

5 minutes per group of 10 (70% accuracy). Obtain 5 digits in each product.

Group A		Group B	
(306)	$0.912 \times 2547 =$	(316)	$0.835 \times 16353 =$
(307)	$2.97 \times 52080 =$	(317)	$0.032 \times 54964 =$
(308)	$0.950 \times 9832 =$	(318)	$0.283 \times 4046 =$
(309)	$87 \times 0.98993 =$	(319)	$93 \times 0.61746 =$

APPENDIX A

Group A

(310)　$4.7 \times 92825 =$
(311)　$5.15 \times 9606 =$
(312)　$0.88 \times 50448 =$
(313)　$0.083 \times 20424 =$
(314)　$171 \times 0.98608 =$
(315)　$0.556 \times 40419 =$

Group B

(320)　$977 \times 2.2020 =$
(321)　$8.3 \times 64565 =$
(322)　$14.0 \times 35581 =$
(323)　$4.15 \times 3461 =$
(324)　$742 \times 5.4151 =$
(325)　$1.33 \times 29095 =$

APPENDIX B

Answers to Exercises of Appendix A

Addition

(1)	774	(6)	17165
(2)	1451	(7)	9431
(3)	1607	(8)	8412
(4)	7588	(9)	5913
(5)	8271	(10)	14861

(A) Addition of 2-digit numbers

(11)	629
(12)	623
(13)	486
(14)	484
(15)	568

(B) Addition of 3-digit numbers

(16)	6960
(17)	6090
(18)	4410
(19)	6185
(20)	5025

(C) Addition of 4-digit numbers

(21)	42896	(26)	48762
(22)	56344	(27)	52741
(23)	56307	(28)	50248
(24)	40628	(29)	44596
(25)	68071	(30)	52734

(D) Addition of 5-digit numbers

(31)	525865	(36)	517168
(32)	537601	(37)	610962
(33)	647469	(38)	419533
(34)	474782	(39)	487431
(35)	518718	(40)	546349

Subtraction

(41)	18593	(46)	72616
(42)	11335	(47)	18579
(43)	69819	(48)	25211
(44)	17212	(49)	15038
(45)	25243	(50)	23076

Combined Addition and Subtraction

(51)	61973
(52)	178735
(53)	24863
(54)	64652
(55)	296792

Multiplication

(A) Multiplication of 2-digit numbers

(56)	5428	(61)	2772
(57)	4042	(62)	2158
(58)	6319	(63)	1602
(59)	3900	(64)	2134
(60)	828	(65)	3243

(B) Multiplication of 3-digit numbers

(66)	377454	(71)	154833
(67)	376443	(72)	216825
(68)	468468	(73)	446754
(69)	123888	(74)	579700
(70)	284273	(75)	152218

170   THE ABACUS

(C) Multiplication of 4-digit numbers

        (76)   1277   (81)   511.5
        (77)   727.0   (82)   885.9
        (78)   481.2   (83)   25760
        (79)   5748   (84)   16220
        (80)   3255   (85)   358.7

(D) Multiplication of 5-digit numbers

        (86)   18685   (91)   21.052
        (87)   8604.6   (92)   41560
        (88)   62.424   (93)   584290
        (89)   152.56   (94)   5739.7
        (90)   11.592   (95)   3.3143

Division (A)

        (96)   76537   (101)   89
        (97)   8179   (102)   75
        (98)   58608   (103)   817
        (99)   92   (104)   473
        (100) 47   (105)   187

Division (B)

        (106) 618   (116)   116
        (107) 938   (117)   4.32
        (108) 73900   (118)   381
        (109) 194   (119)   120
        (110) 11100   (120)   4120
        (111) 2960   (121)   445
        (112) 818   (122)   173
        (113) 79900   (123)   3610
        (114) 223   (124)   1110
        (115) 40.8   (125)   53.9

Division (C)

(126)	2.2981	(136)	619.70
(127)	0.45106	(137)	1.1953
(128)	2.0461	(138)	2.7325
(129)	0.80916	(139)	2725.1
(130)	2.0227	(140)	0.36697
(131)	4.9878	(141)	0.65291
(132)	0.51200	(142)	4.1842
(133)	0.78634	(143)	0.63214
(134)	0.17023	(144)	405.41
(135)	0.14437	(145)	936.32

Addition

6th-Grade Operator

(146)	7621	(156)	13755
(147)	8033	(157)	18734
(148)	4774	(158)	10133
(149)	8455	(159)	13705
(150)	4260	(160)	17706
(151)	8151	(161)	20216
(152)	7003	(162)	7990
(153)	9521	(163)	13022
(154)	2884	(164)	7674
(155)	8009	(165)	12252

5th-Grade Operator

(166)	26708	(181)	49191
(167)	30659	(182)	31427
(168)	40869	(183)	61467
(169)	28530	(184)	47995
(170)	24650	(185)	61182
(171)	42382	(186)	28205
(172)	22115	(187)	27666
(173)	32328	(188)	40891
(174)	25239	(189)	21484
(175)	31964	(190)	33638

(176)	52171	(191)	36283
(177)	28402	(192)	20204
(178)	59320	(193)	23410
(179)	45837	(194)	15370
(180)	54642	(195)	23455

4th-Grade Operator

(196)	81299	(211)	344335
(197)	41397	(212)	259622
(198)	77777	(213)	239407
(199)	59526	(214)	183202
(200)	83868	(215)	387134
(201)	67991	(216)	225226
(202)	62160	(217)	312832
(203)	75992	(218)	279112
(204)	52198	(219)	359453
(205)	66179	(220)	310822
(206)	422854	(221)	249210
(207)	213876	(222)	122656
(208)	241032	(223)	274807
(209)	143124	(224)	175403
(210)	358966	(225)	249590

3rd-Grade Operator

(226)	2229273	(236)	1011678
(227)	2158731	(237)	857369
(228)	2363860	(238)	711638
(229)	1501564	(239)	780297
(230)	2814820	(240)	641316
(231)	1973044	(241)	629867
(232)	1879700	(242)	774537
(233)	2262130	(243)	940362
(234)	2313736	(244)	898869
(235)	1225695	(245)	896545

Multiplication

6th-Grade Operator

(246)	83904	(256)	249256
(247)	61838	(257)	134790
(248)	34830	(258)	343989
(249)	13770	(259)	33028
(250)	60074	(260)	129344
(251)	8064	(261)	204138
(252)	11622	(262)	416883
(253)	33696	(263)	91562
(254)	11600	(264)	300330
(255)	15466	(265)	511575

5th-Grade Operator

(266)	238680	(276)	289080
(267)	166166	(277)	162745
(268)	70070	(278)	533676
(269)	187668	(279)	218078
(270)	251247	(280)	136136
(271)	429338	(281)	177995
(272)	140580	(282)	75504
(273)	174286	(283)	276165
(274)	496910	(284)	205997
(275)	412508	(285)	756960

4th-Grade Operator

(286)	4741225	(296)	1868856
(287)	3199353	(297)	988676
(288)	5449356	(298)	6038565
(289)	5156228	(299)	4533408
(290)	3143002	(300)	1674636
(291)	2783826	(301)	820204
(292)	2331984	(302)	8914257
(293)	950544	(303)	4471296
(294)	1611938	(304)	4786361
(295)	4876053	(305)	1419348

3rd-Grade Operator

(306)	2322.9	(316)	13655
(307)	154680	(317)	1758.8
(308)	9340.4	(318)	1145.0
(309)	86.124	(319)	57.424
(310)	436280	(320)	2151.4
(311)	49471	(312)	535890
(312)	44394	(322)	498130
(313)	1695.2	(323)	14363
(314)	168.62	(324)	4018.0
(315)	22473	(325)	38696

Acknowledgments

It is a pleasure to thank those who allowed us to print the quotations given at the beginning of this book: Collier's Encyclopedia for the quotation by Hartley Howe, The Royal Society of Edinburgh for the quotation by C. G. Knott, and Dover Publications for the quotation by Martin Gardner.

I am indebted also to various manufacturers for photographs. Keuffel and Esser kindly furnished the slide-rule picture, Fig. 1.1. Smith-Corona Marchant, and the Friden Division of the Singer Company are responsible for Figs. 1.2 and 1.3. And the International Business Machines Corporation should be thanked for Fig. 1.4.

Index

Abacus, Chinese   30
  Japanese   30
  Roman   29
  Russian   30
  minimal   76
Abacus, $(0, m)$ 78, 101
  (n, m) 82, 92, 126
  (p, n, m) 96, 141
Abacus, $(0, 5)$ 119
  $(0, 7)$ 122
  $(0, 9)$ 78, 101
  $(0, 10)$ 78
  $(0, 11)$ 122
  $(0, 14)$ 79
  $(1, 4)$ 80
  $(1, 5)$ 84
  $(2, 5)$ 85, 126
  $(10, 10)$ 139
  $(2, 2, 4)$ 86, 141
  $(-4, 2, 4)$ 88, 145
Abacus Association of America   4
ADAMS, JOHN QUINCY   12, 20
Addition   40
  on Greek board   23
  on European board   27
  on Japanese abacus   32
  on $(0, 9)$ abacus   102
  on $(2, 5)$ abacus   127
  on $(2, 6)$ abacus   135
  on $(2, 2, 4)$ abacus   141
  on $(-4, 2, 4)$ abacus   146
Addition tables   40, 41, 42
Additive parts   62
Almagest   20
ANDREWS, F.E.   20, 73, 100, 140
Answers   168
Approximation   113

Arithmetic   40
Arithmos   11
Attic numerals   6

Babylonian numerals   7
BARNARD, F.P.   39, 100
Base, number   5, 11
Base-2 system   13
  3   13
  6   119
  8   14, 122
  12   15, 17, 122
  60   139
Basic method of multiplication   106
Beads   74, 90
BLATER, J.   73
BROWN, J.F.   72

CAJORI, F.   20
Calculator, desk   2
  electronic   2
CASSINA, U.   72
Chinese abacus   30
Chinese numerals   7
COLLES, G.W.   20
COLLIER'S ENCYCLOPEDIA   39
Complements   42
Computer, sequential   2
Condensed multiplication tables   46
COOLIDGE, J.L.   73, 125
Counters   74, 90
Counting   5
Counting boards, Greek   21
  European   24
CRELLE, A.L.   73
Criteria of abacus excellence   81
CUTLER, A.   73, 125

177

Design 74
Desk calculator 2
Digits 6, 11
Dimensions of beads 90
Division 65
  on Japanese abacus 37
  on (0, 9) 115
  on (2, 5) 132
  on (2, 6) 137
DOMORYAD, A.P. 73, 125, 140
Doubling and halving 106, 108
Duodecimal system (Base-12) 15, 17
Duplation and mediation 28, 54

Egyption numerals 6
ELBROW, G. 20
Electronic calculator 2
ESSIG, J. 20, 100
European counting board 24
Exercises, addition 150
  subtraction 152
  multiplication 153
  division 154

Factorials 15
Factors 17, 62, 64, 66, 71

GALLE, A. 19, 125
Gelosia method 57
Gobar numerals 8
Greek numerals (500 B.C.) 9

Hebrew numerals 9
Hexagesimal system (Base-60) 19
Hindu numerals 7
HORSBURGH, E.M. 19, 73, 125

INGALLS, W.R. 20
Integers 10
International Abacus Association 4
Irregular number 66
IRWIN, W.C. 19

Japan Chamber of Commerce and Industry
  4, 20, 100, 125

Japanese abacus 30
Jetons 25

KNOTT, C.G. 39
KOJIMA, T. 20, 39, 100, 125

LEECH, T. 20
LESLIE, J. 20, 45, 73, 100
LEYBOURN, W. 73

MC SHANE, R. 73, 125
Mayan numerals 8
MENNINGER, K. 20, 38, 39, 73, 100, 125, 140
Meros 12
Metric system 17
MEYER ZUR CAPELLEN, W. 19
MEYERS, L. 72
MIKAMI, Y. 39
Millionaire calculator 125
Minimal abacus 76, 91
MOON, P. 20
Multiplication 55
  on European board 28
  on (0, 9) abacus 104
  on (2, 5) 129
  on (2, 6) 135
  on (2, 2, 4) 143
  on (−4, 2, 4) 148
Multiplication table 44
MURRAY, F.J. 19, 125

NAGL, A. 39
NAPIER, J. 73, 125
Napier's rods 58, 106, 111, 131, 136
NEUGEBAUER, O. 20, 73, 125
Non-minimal abaci 94
Number 1, 10
  regular 66
  reciprocal 70
Number bases 5, 11
Number of products in multiplication table
  41, 47, 50
Numerals 6, 7, 8, 9

Operating the Japanese abacus 32

Operator, 6th grade, addition 155
  5th grade 157
  4th grade 160
  3rd grade 163
  6th grade, multiplication 165
ORE, O. 20, 73, 100
Other bases 118, 134

Pedagogical value of abacus 5
PETERS, J. 73
Positional notation 9
Problems, addition 150, 155
  subtraction 152
  multiplication 153
  division 154
PTOLEMY, C. 11, 20

Quarter squares 46

Rand Corporation 150
Real numbers 10
Reciprocals 62, 64, 66, 118, 123
Reciprocals of regular numbers, Base-10 70
  Base-12 138
Regular number 66
REISCH, G. 24
RICHARDS, R.K. 19
Roman abacus 29
Roman numerals 6, 10
Russian abacus 30

Salamis counting board 21
SEEBY, S.M. 100
Sequential computer 2
Sexagesimal system (Base-60) 19
SHOCKLEY, J.E. 73
Shortcuts in multiplication 61

Slide rule 1
SMITH, D.E. 19, 20, 38, 39, 72, 73, 100, 125, 140
SNELLING, T. 39
Soroban 20, 30
Specification of abaci 77
SPENCER, HERBERT 18, 20
Suan-pan 30
Subtraction 43
  on Greek board 24
  on European board 28
  on Japanese abacus 34
  on (0, 9) 104
  on (2, 5) 129

TANI, Y. 20, 39, 125
TERRY, G.S. 20, 73, 100, 140
Theory, minimal abaci 91
  non-minimal abaci 94
THUREAU-DANGIN, F. 73
Trachtenberg method 56, 62, 73

Unique setting 76
Unit counters 77

VOISIN, A. 73
VOLPEL, M.C. 72

WAERDEN, B.L. VAN DER 20, 38
WEAST, R.C. 100
Weight 77
WILLERDING, M.F. 73
WILLERS, F.A. 19, 125

YELDHAM, F. 39
YOSHINO, Y. 20, 39, 100, 125

ZIMMERMAN, H. 73

PENSACOLA JUNIOR COLLEGE LIBRARY
QA75 .M64
Moon, Parry Hiram, 1898-   000
The abacus: its history,   210101

3 5101 00090021 2

DATE

DATE DUE
02 24 '02

5 12 99

QA   73-19149
75
.M64

Moon, Paryy Hiram
The abacus: its history, its design,
  its possibilities in the modern
  world

WITHDRAWN

PENSACOLA JUNIOR COLLEGE LIBRARY